Anaerobic treatment and resource recovery from methanol rich waste gases and wastewaters

Tejaswini Eregowda

Joint PhD degree in Environmental Technology

Docteur de l'Université Paris-Est
Spécialité : Science et Technique de l'Environnement

Dottore di Ricerca in Tecnologie Ambientali

Degree of Doctor in Environmental Technology

Tampere University

Thesis for the degree of Doctor of Philosophy in Environmental Technology

PhD thesis – Proefschrift – Väitöskirja – Tesi di Dottorato – Thèse

Tejaswini EREGOWDA

Anaerobic treatment and resource recovery from methanol rich waste gases and wastewaters

To be defended on 23 May, 2019

In front of the PhD committee

Prof. Bo H. Svensson	Chairperson, Reviewer
Prof. Largus Angenent	Reviewer
Dr. Tanja Radu	Reviewer
Prof. Piet N. L. Lens	Promotor
Prof. Jukka Rintala	Co-promotor
Prof. Giovanni Esposito	Co-promotor
Prof. Eric D. van Hullebusch	Co-promotor
Asst. Prof. Marika Kokko	Co-promoter

Marie Skłodowska-Curie European Joint Doctorate,
Advance biological waste-to-energy technologies
(ABWET)

Evaluation committee:

Chairperson

Prof. Bo H. Svensson
Professor, Department of thematic studies, Environmental Change
Linköping University, Linköping, Sweden

Reviewers

Prof. Largus T. Angenent
Professor of Environmental Biotechnology
University of Tübingen, Germany

Prof. Bo H. Svensson
Professor, Department of thematic studies, Environmental Change
Linköping University, Linköping, Sweden

Dr. Tanja Radu
Lecturer in Water Engineering, School of Architecture, Building and Civil Engineering
Loughborough University, Leicestershire, UK

Thesis Promotor

Prof. Piet N. L. Lens
Department of Environmental Engineering and Water Technology,
IHE Delft Institute for Water Education, The Netherlands

Thesis Co-Promotors

Prof. Jukka Rintala
Faculty of Engineering and Natural Sciences, Tampere University, Finland

Prof. Giovanni Esposito
Department of Civil and Mechanical Engineering,
University of Cassino and Southern Lazio, Italy

Prof. Eric D. van Hullebusch
University of Paris-Est, France

Asst. Prof. Marika E. Kokko
Faculty of Engineering and Natural Sciences, Tampere University, Finland

Supervisory team

Thesis Supervisor

Prof. Piet N. L. Lens
Department of Environmental Engineering and Water Technology,
IHE Delft Institute for Water Education, The Netherlands

Thesis Co-Supervisor

Prof. Jukka Rintala
Faculty of Engineering and Natural Sciences, Tampere University, Finland

Thesis mentors

Dr. Eldon R. Rene
Department of Environmental Engineering and Water Technology
IHE Delft, Institute for Water Education, Delft, The Netherlands

Asst. Prof. Marika E. Kokko
Faculty of Engineering and Natural Sciences, Tampere University, Finland

This research was conducted in the framework of the Marie Sklodowska-Curie European Joint Doctorate (EJD) in Advanced Biological Waste-to-Energy Technologies (ABWET) and supported by Horizon 2020 under the grant agreement no. 643071. This research was conducted under the auspicious of the Graduate School for Socio-Economic and Natural Sciences of the Environment (SENSE).

CRC Press/Balkema is an imprint of the Taylor & Francis Group, an informa business
© 2019, Tejaswini Eregowda

Although all care is taken to ensure integrity and the quality of this publication and the information herein, no responsibility is assumed by the publishers, the author nor IHE Delft for any damage to the property or persons as a result of operation or use of this publication and/or the information contained herein.

A pdf version of this work will be made available as Open Access via http://repository.tudelft.nl/ihe

This version is licensed under the Creative Commons Attribution-Non Commercial 4.0 International License, http://creativecommons.org/licenses/by-nc/4.0/

Published by:
CRC Press/Balkema
Schipholweg 107C, 2316 XC, Leiden, the Netherlands
Pub.NL@taylorandfrancis.com
www.crcpress.com – www.taylorandfrancis.com
ISBN: 978-0-367-41846-5

Table of Contents

List of abbreviations

ANME	Anaerobic methanotrophs
AOM	Anaerobic oxidation of methane
BF	Biofilter
BTF	Biotrickling filter
COD	Chemical oxygen demand
DAMO	Denitrifying anaerobic methane oxidation
EBRT	Empty bed residence time
EC	Elimination capacity
EC_{max}	Maximum elimination capacity
FC	Foul condensate
FID	Flame ionization detector
GC-MS	Gas chromatography-mass spectrometry
GHG	Greenhouse gases
GWP	Global warming potential
HRT	Hydraulic retention time
ILR	Inlet loading rate
MBR	Membrane bioreactor
MDH	Methanol dehydrogenase
$MtCO_2e$	Metric tons of carbon dioxide equivalent
NCG	Non-condensable gases
ppb	Parts per billion
ppm	Parts per million
RS	Residence time
RT	Retention time
RuMP	Ribulose mono phosphate
SMWW	Synthetic methanol-rich wastewater
SRB	Sulfate reducing bacteria
UASB	Upflow anaerobic sludge blanket
VFA	Volatile fatty acids
VIC	Volatile inorganic carbon
VOC	Volatile organic carbon

Summary

Methanol is an important volatile organic compound (VOC) present in the gaseous and liquid effluents of process industries such as pulp and paper, paint manufacturing and petroleum refineries. An estimated 55,377 tonnes of methanol was emitted to the atmosphere in the year 2017 in the United States alone and at least 65% of the total emission was from the Kraft mills of the pulp and paper industries. The anaerobic biological treatment of methanol-rich gaseous and liquid effluents was tested in two bioreactor configurations, namely a biotrickling filter (BTF) and an upflow anaerobic sludge blanket (UASB) reactor. The volatile fatty acids (VFA) produced in these bioreactors were quantified and a mass balance analysis was carried out.

Gas-phase methanol removal along with thiosulfate (~ 1000 mg/L) reduction was carried out for 123 d in an anoxic BTF. A maximum elimination capacity (EC_{max}) of 21 g/m^3.h for methanol and complete removal of thiosulfate was achieved. To examine the gas-phase methanol removal along with selenate reduction, another anoxic BTF was operated for 89 d under step and continuous selenate feeding, wherein the selenate removal efficiency was > 90% and ~ 68%, respectively, during step and continuous selenate feed and a methanol EC_{max} of 46 g/m^3.h was achieved. The anaerobic bioreduction of selenate coupled to methane oxidation was investigated in batch reactors and a BTF inoculated with marine sediment and operated for a period of 348 d. Complete reduction of up to 140 mg/L of step fed selenate was achieved in the BTF. Furthermore, the effect of selenate, sulfate and thiosulfate on methanol utilization for VFA production was individually examined in batch systems.

For the study on liquid-phase methanol, acetogenesis of foul condensate (FC) obtained from a chemical pulping industry was tested in three UASB reactors operated at 22, 37 and 55 °C for 51 d. A maximum methanol removal efficiency of 45% in the 55 °C reactor and nearly complete removal of ethanol and acetone in all UASB reactors was achieved. Prior to acetogenesis of the FC, the UASB reactors were operated for a period of 113 d under conditions reported to induce acetogenesis of methanol-rich synthetic wastewater.

The recovery of VFA was explored through adsorption studies using anion exchange resins in batch systems. The trends and capacity of adsorption of individual VFA on Amberlite IRA-67 and Dowex optipore L-493 were examined by fitting the experimental data to adsorption isotherm and kinetic models. Subsequently, a sequential batch process was tested to achieve selective separation of acetate from the VFA mixture.

Samenvatting

Methanol is een belangrijke vluchtige organische stof aanwezig in de gasvormige en vloeibare effluenten van procesindustrieën zoals pulp en papier, verfproductie en petroleum raffinaderijen. Naar schatting werd in 2017, 55.377 ton methanol in de atmosfeer in de uitgestoten Verenigde Staten alleen en ten minste 65% van de totale uitstoot was afkomstig van de Kraft-fabrieken in de pulp- en papierindustrie. De anaërobe biologische behandeling van methanolrijke gasvormige en vloeibare effluenten werd getest in bioreactorconfiguraties zoals biotrickling filters (BTF) en upflow anaerobe slibdeken (UASB) reactoren. De in deze bioreactoren geproduceerde vluchtige vetzuren (VFA) werden gekwantificeerd en er werd een massa balans analyse uitgevoerd.

Gasfase methanol verwijdering samen met thiosulfaat (~ 1000 mg/L) reductie werd uitgevoerd voor 123 dagen in een anoxische BTF. Een maximale eliminatiecapaciteit (EC_{max}) van 21 $g/m^3.h$ voor methanol en volledige verwijdering van thiosulfaat werd bereikt. Om de gasfase-methanolverwijdering te onderzoeken samen met selenaat reductie, werd nog een anoxische BTF gedurende 89 dagen onder stap en continue selenaatvoeding bedreven. De verwijdering van het selenaat was > 90% en ~ 68% tijdens, respectievelijk, stap en continue toevoer van selenaat en een EC_{max} van methanol van 46 $g/m^3.h$ werd bereikt. Anaërobe bioreductie van selenaat gekoppeld aan methaanoxidatie werd onderzocht in batch reactoren en een BTF geïnoculeerd met mariene sedimenten gedurende 348 d. Volledige reductie van stapgevoed selenaat (tot 140 mg/L) werd bereikt. Verder werden de effecten van verschillende concentraties van selenaat, sulfaat en thiosulfaat, individueel, op het gebruik van methanol voor VFA-productie onderzocht in batch systemen.

Voor de studie naar methanol in de vloeistoffase werd acetogenese van vervuild condensaat van een chemische pulpindustrie getest in drie UASB-reactoren die gedurende 51 d bij 22, 37 en 55 °C werden bedreven. Een maximum methanolverwijderingsrendement van 45% in de 55 °C-reactor en bijna volledige verwijdering van ethanol en aceton werd bereikt in alle UASB-reactoren. Voorafgaand aan de acetogenese van het condensaat werden de UASB-reactoren gedurende een periode van 113 dagen in bedrijf gesteld onder omstandigheden waarvan is gerapporteerd dat zij acetogenese van methanolrijk synthetisch afvalwater induceren.

Het terugwinnen van VFA werd onderzocht door middel van adsorptieonderzoeken met behulp van anionenwisselingsharsen in batchsystemen. De adsorptiecapaciteit van individuele VFA op Amberlite IRA-67 en Dowex optipore L-493 werd onderzocht door de experimentele gegevens te fitten met adsorptie-isothermen en kinetische modellen. Vervolgens werd een sequentieel batchproces getest om selectieve scheiding van acetaat uit het VFA-mengsel te bereiken.

Yhteenveto

Metanoli on tärkeä haihtuva orgaaninen yhdiste, jota on sekä kaasumaisissa että nestemäisissä sivu- ja jätevirroissa, joita tuotetaan prosessiteollisuudessa kuten massa- ja paperiteollisuudessa, maaliteollisuudessa sekä öljynjalostuksessa. Vuonna 2017 ilmakehään pääsi pelkästään Yhdysvalloissa noin 55377 tonnia metanolia, josta vähintään 65% oli peräisin sellu- ja paperiteollisuuden sulfaattiprosessista. Tässä tutkimuksessa metanolia sisältäviä kaasumaisia ja nestemäisiä virtoja käsiteltiin biologisesti anaerobisissa bioreaktoreissa, kuten suodinreaktorissa (BTF) ja lietepatjareaktorissa (UASB). Näissä bioreaktoreissa metanolista tuotetut haihtuvat rasvahapot määritettiin kvantitatiivisesti ja reaktoreiden massatasapainot selvitettiin.

Metanolia poistettiin kaasufaasista pelkistämällä samalla tiosulfaattia (~ 1000 mg/L) anaerobisessa biosuotimessa, jota ajettiin 123 päivää. Kokeissa saavutettiin metanolin maksimipoistotehokkuudeksi 21 g/m^3.h, kun taas tiosulfaatti poistettiin täydellisesti. Kaasufaasin metanolin poiston yhdistämistä selenaatin pelkistykseen tutkittiin toisessa anaerobisessa biosuotimessa 89 päivää syöttämällä metanolia ja selenaattia reaktoriin joko panos- tai jatkuvatoimisesti. Selenaatin poistotehokkuus oli > 90% panostoimisessa ja noin 68% jatkuvatoimisessa syötössä. Metanolin maksimipoistotehokkuus oli 46 g/m^3.h. Metanolin poistoa yhdistettynä selenaatin pelkistykseen tutkittiin edelleen panostoimisessa anaerobisessa biosuotimessa, johon rikastettiin mikrobiyhteistö merisedimentistä. Reaktoria operoitiin 348 päivää, jonka aikana saavutettiin täydellinen selenaatin pelkistys jopa 140 mg/L selenaatin konsentraatiolla. Reaktorikokeiden lisäksi selenaatin, sulfaatin ja tiosulfaatin eri konsentraatioiden vaikutuksia haihtuvien rasvahappojen tuottoon metanolista tutkittiin panossysteemeissä.

Sellutehtaan nestemäistä metanolia sisältävää lauhdetta käsiteltiin kolmessa lietepatjareaktorissa, joita operoitiin 22, 37 ja 55 °C:ssa 51 päivää. Maksimi metanolinpoistotehokkuus oli 45% 55°C:ssa operoidussa reaktorissa ja lähes 100% etanolin ja asetonin poisto saavutettiin kaikissa UASB-reaktoreissa. Ennen lauhteen käsittelyä UASB-reaktorissa, reaktoria operoitiin 113 päivää synteettisellä metanolia sisältävällä jätevedellä olosuhteissa, jotka edesauttavat metanolin asetogeneesiä haihtuviksi rasvahapoiksi.

Haihtuvien rasvahappojen talteenottoa anioninvaihtoadsorbenteilla tutkittiin panoskokeissa. Yksittäisten haihtuvien rasvahappojen adsorptiokapasiteettia tutkittiin Amberlite IRA-67 ja Dowex optipore L-493 adsorbenteilla sovittamalla kokeelliset tulokset adsorptio-isotermiin ja kineettisiin malleihin. Lopuksi testattiin asetaatin selektiivistä erottamista haihtuvia rasvahappoja sisältävästä liuosfaasista peräkkäisellä panosprosessilla.

Sommario

Il metanolo è un importante composto organico volatile (VOC) presente in effluenti liquidi e gassosi di industrie come cartiere, produttrici di vernici e raffinerie. Una quantità stimata di 55,377 ton di metanolo è stata emessa in atmosfera nel 2017 negli Stati Uniti, 65% della quale proveniente dal processo *Kraft* delle cartiere. Il trattamento biologico degli effluenti liquidi e gassosi ricchi in metanolo è stato testato in bioreattori a letto percolatore (BTF) e a letto di fango a flusso verticale (UASB). Gli acidi grassi volatili (VFA) prodotti in questi bioreattori sono stati quantificati per effettuare bilanci di massa.

La rimozione di metanolo gassoso tramite riduzione del tiosolfato (~ 1000 mg/L) è stata effettuate in un BTF anossico per 123 giorni, ottenendo una capacità massima di eliminazione (EC_{max}) del metanolo di 21 $g/m^3.h$ e una rimozione completa del tiosolfato. Per studiare la rimozione del metanolo tramite riduzione del selenato, un altro BTF è stato utilizzato per 89 giorni con alimentazione di selenato sia intermittente sia continua. L'efficienza di rimozione del selenato è stata > 90% e ~ 68% con l'alimentazione intermittente e continua, rispettivamente, ottenendo una EC_{max} del metanolo di 46 $g/m^3.h$. La bioriduzione anaerobica del selenato in coppia con l'ossidazione del metano è stata investigata per 348 giorni in reattori batch e in un BTF inoculati con sedimento marittimo, ottenendo una rimozione completa del selenato, alimentato ad intermittenza fino a 140 mg/L. Inoltre, l'effetto della concentrazione di selenato, solfato e tiosolfato, individualmente, sulla produzione di VFA da metanolo è stato esaminato in batch.

Per gli esperimenti sul metanolo in fase liquida, l'acetogenesi da condensati contaminati (FC), ottenuti da un processo di spappolamento chimico, è stata studiata per 51 giorni in tre reattori UASB a 22, 37 e 55 gradi. Un'efficienza massima di rimozione del 45% è stata ottenuta a 55 °C, mentre una quasi completa rimozione di etanolo ed acetone è stata ottenuta in tutti e tre gli UASB. Prima di essere usati per acetogenesi da FC, i reattori UASB erano stati adattati per 113 giorni sotto condizioni note da indurre acetogenesi da acque sintetiche ricche in metanolo.

Il recupero di VFA è stato testato in batch mediante studi di adsorbimento in resine a scambio anionico. La capacità di adsorbimento dei singoli VFA sull'Amberlite IRA-67 e la Dowex optipore L-493 è stata esaminata tramite interpolazione dei dati sperimentali sulle isoterme di adsorbimento e i modelli cinetici. In seguito, un processo in batch è stato testato per ottenere la separazione selettiva dell'acetato dalla miscela di VFA.

Résumé

Le méthanol est un composé organique volatil (COV) important, présent dans les effluents gazeux et liquides des industries de transformation telles que les pâtes à papier et papiers, la fabrication de peinture et les raffineries de pétrole. Aux États-Unis seulement, environ 55 377 tonnes de méthanol ont été émises dans l'atmosphère en 2017 et au moins 65% du total des émissions provenaient des usines Kraft des industries de la pâte à papier et du papier. Le traitement biologique anaérobie des effluents gazeux et liquides riches en méthanol a été testé dans des configurations de bioréacteurs, telles que des filtres à biotrickling (BTF) et des réacteurs à lit de boues anaérobie à flux ascendant (UASB). Les acides gras volatils (AGV) produits dans ces bioréacteurs ont été quantifiés et une analyse de bilan massique a été réalisée.

Une élimination du méthanol en phase gazeuse ainsi qu'une réduction du thiosulfate (~ 1000 mg/L) ont été effectuées pendant 123 jours dans un BTF anoxique. Une capacité d'élimination maximale (EC_{max}) de 21 g/m^3.h pour le méthanol et l'élimination complète du thiosulfate ont été atteintes. Pour examiner l'élimination du méthanol en phase gazeuse ainsi que la réduction du sélénate, un autre BTF anoxique a été mis en œuvre pendant 89 jours sous une alimentation par étapes et continue en sélénate. L'efficacité d'élimination du sélénate était respectivement > 90% et ~ 68% au cours d'une alimentation par étapes et continue en sélénate et une EC_{max} de méthanol de 46 g/m^3.h. La bioréduction anaérobie du sélénate couplée à l'oxydation du méthane a été étudiée dans des réacteurs discontinus et un BTF inoculé avec un sédiment marin et opéré pendant 348 jours et une réduction complète pouvant atteindre 140 mg/L de sélénate alimenté par étapes a été obtenue. En outre, les effets de différentes concentrations de sélénate, de sulfate et de thiosulfate, individuellement, sur l'utilisation du méthanol pour la production d'AGV ont été examinés dans des systèmes de traitement par lots.

Pour l'étude sur le méthanol en phase liquide, l'acétogénèse du condensat encrassé (FC) obtenu dans une industrie de réduction en pâte chimique a été testée dans trois réacteurs UASB fonctionnant à 22, 37 et 55 °C pendant 51 jours. Une efficacité d'élimination maximale du méthanol de 45% dans le réacteur à 55 °C et une élimination presque complète de l'éthanol et de l'acétone ont été obtenues dans tous les réacteurs UASB. Avant l'acétogénèse du FC, les réacteurs UASB ont été mis en fonctionnement pendant 113 jours dans des conditions rapportées induir une acétogenèse des eaux usées synthétiques riches en méthanol.

La récupération d'AGV a été explorée au moyen d'études d'adsorption utilisant des résines échangeuses d'anions dans des systèmes de traitement par lots. La capacité d'adsorption de différents AGV sur Amberlite IRA-67 et Dowex optipore L-493 a été examinée en ajustant les données expérimentales à des modèles isothermes et cinétiques d'adsorption. Ensuite, un processus séquentiel par lots a été testé pour obtenir une séparation sélective de l'acétate du mélange d'AGV.

Acknowledgements

This work would not be possible without the support by Marie Skłodowska-Curie European Joint Doctorate (EJD) in Advanced Biological Waste-To Energy Technologies (ABWET) funded by the Horizon 2020 program under the grant agreement no. 643071.

Just like a bioconversion that would not function optimally without the addition of micronutrients, this PhD thesis would not be complete without the professional and moral guidance and support from several people during its course.

I sincerely thank Prof. Piet N L Lens, promotor of this thesis, for his constant supervision, encouragement and critics that allowed me to professionally grow as a researcher. His guidance and comments were vital for the success of this research. My earnest thanks to Prof. Jukka Rintala, co-promotor of this thesis, for his support, encouragement and insightfulness towards my research during my mobility at TUT, Finland. You are a true inspiration. Thanks to Prof, Giovanni Esposito and Prof. Eric van Hullebusch for their thoughtful comments and suggestions during the summer schools and support with the administrative affairs. Dr. Eldon R Rene has been a terrific mentor, constantly available and supportive about every pitfall. His guidance towards improving my knowledge, communication, critical thinking were very essential and I cannot thank him enough for his positive attitude. Thank you Asst. Prof. Marika Kokko for the guidance and support during the mobility in Finland. I appreciate the help extended by Iosif Scoullos, Paolo Dessi, Marika Kokko towards summary translation and Sheba Nair for turnitin handling. I thank Dr. Pritha Chatterjee for the guidance and analytical support with the lab work at TUT, the lab staff of IHE Delft and TUT for the support with lab procurements. Thank you Dr. Jack van de Vossenberg for being supportive during the initial days of my PhD.

A big thanks to, Iosif, Pritha, Chris, Gabriele, Shrutika, Mohammed, Paolo, Viviana, Samayita, Joseph, Raghu, Georg, Lea, Angelo, Mirko, Karan, Sheba, Ramita and the ABWET PhDs for being there to make this long journey less difficult. I am glad our paths crossed and I will truly treasure having known beautiful souls such as you all. Thank you Bhartish for always being there. Your company has been a great comfort throughout.

Lastly, I thank my family who have constantly stood by through every walk that has led me here. None of this would be possible without their love, support, criticism and encouragement. Especially from my mother and sister.

Author's contribution

Paper I: Tejaswini Eregowda set up the reactor, Luck Matanhike and Tejaswini Eregowda operated the reactor and carried out the data analysis during the operational period 1-90 d. 91-123 d of the reactor operation and data analysis was solely carried out by Tejaswini Eregowda. Tejaswini Eregowda wrote the manuscript. Luck Matanhike, Eldon Rene and Piet Lens participated in planning the experiments, helped in data interpretation and thoroughly revised the manuscript.

Paper II: Tejaswini Eregowda performed all the experiments, data analysis and wrote the manuscript. Eldon Rene and Piet Lens participated in planning the experiments, helped in data interpretation and thoroughly revised the manuscript.

Paper III: Tejaswini Eregowda performed all the experiments, data analysis and wrote the manuscript. Jukka Rintala, Eldon Rene and Piet Lens participated in planning the experiments, helped in data interpretation and thoroughly revised the manuscript.

Paper IV: Tejaswini Eregowda performed the experiments, data analysis and wrote the manuscript. Marika Kokko, Eldon Rene, Jukka Rintala, and Piet Lens participated in planning the experiments, helped in data interpretation and thoroughly revised the manuscript.

Paper V: Tejaswini Eregowda and Luck Matanhike contributed equally for performing the experiments and data analysis. Tejaswini Eregowda wrote the manuscript. Eldon Rene and Piet Lens participated in planning the experiments, helped in data interpretation and thoroughly revised the manuscript.

CHAPTER 1

General introduction

1.1. Background

Methanol is an important volatile organic compound (VOC) present in the gaseous and liquid effluents of process industries such as pulp and paper, paint manufacturing and petroleum refineries. In 2017, the total chemical pulp production accounted for 27.63 million tonnes in the CEPI member countries (Confederation of European Paper Industries) (CEPI, 2017). Given that ~4 kg methanol is produced per tonne pulp produced (Lin, 2008), 0.11 million tonnes of methanol is potentially available as a resource in the CEPI member states alone. An estimated 55,377 tonnes of methanol was emitted to the atmosphere in the year 2017 in the United States alone and at least 65% of the total emission was from the Kraft mills from the pulp and paper industries (TRI Program, 2017). In the Kraft pulping process, the condensates from the black liquor digesters and evaporators are subjected to stripping to reduce the fresh water intake and organic load of the wastewater treatment plant. During the stripping process, methanol and other VOC from the condensates accumulate in the overhead vapour as non-condensable gases (NCG), which are generally incinerated (Siddiqui and Ziauddin, 2011; Suhr et al., 2015).

Similarly, methane is a major constituent of the gaseous emissions from the mining and petroleum based industries. Methane is a rich source of energy and it is a major constituent of the natural gas produced as a by-product during the petroleum extraction and has a global warming potential (GWP) potential of 21 (Howarth et al., 2011). The global anthropogenic methane emissions are projected to increase to 8585 metric tonnes CO_2 equivalents ($MtCO_2e$) by 2030 and up to 40 % of the total emissions would be from the energy sector (coal mining operations, natural gas and oil systems) (EPA, 2012). Venting and inefficient flaring of natural gas produced as a by-product of petroleum extraction also accounted for 140 billion cubic meters of natural gas released into atmosphere worldwide (Elvidge et al., 2018).

Through anaerobic biological treatment of gaseous and liquid industrial effluents rich in methanol or methane, resource recovery is possible by the production of economically important by-products like volatile fatty acids (VFA). VFA have a wide range of applications in the food, pharmaceutical and polymer industries and as energy source for the biological removal of nutrients (such as nitrates and phosphates) from wastewater (Choudhari et al., 2015; Jones et al., 2015; Lee et al., 2014). As a consequence, several recent studies have focused on the production and recovery of waste derived VFA from sewage sludge as well as agricultural and food wastes (Rebecchi et al., 2016; Zhou et al., 2017).

Sulfur oxyanions such as sulfate and thiosulfate are common constituents of the whitewater from the pulp and paper and mining industries, tanneries and petroleum refineries (Meyer and Edwards, 2014a; Pol et al., 1998). Selenium (Se) is a trace element that is chalcophillic in nature (strong affinity with sulfur) and Se compounds have a close anology with sulfur compounds (Mehdi et al., 2013; Tan et al., 2016). Effluents from the flue gas desulfurization and copper, sulfur and gold mining industries are rich in Se oxyanions that needs to be treated before discharge to the environment (Tan et al., 2016). Methanol and methane rich liquid and gaseous effluents can potentially be used as the electron donors for the treatment of Se and S oxyanion rich industrial effluents as described by the stoichiometric equations in ***Table 1.1***.

Table 1.1: *Equations for sulfate, thiosulfate and selenate reduction using methanol as electron donor and selenate reduction using methane as electron donor and acetogenesis of methanol.*

Thiosulfate and sulfate reduction and thiosulfate disproportionation using methanol as electron donor:
$S_2O_3^{2-} + H_2O \rightarrow SO_4^{2-} + HS^- + H^+$
$CH_3OH + S_2O_3^{2-} + H^+ \rightarrow HCO3^- + 2HS^- + H_2O$
$4CH_3OH + 3SO_4^{2-} \rightarrow 4\ HCO_3^- + 3HS^- + 4H_2O + H^+$
$CH_3COO^- + S_2O_3^{2-} + H_2O \rightarrow 2HCO^{3-} + 2HS^- + 2H^+$
$CH_3COO^- + SO_4^{2-} \rightarrow 2HCO^{3-} + HS^-$
Selenate reduction with methanol or methane as electron donor
$4CH_3OH + 3SeO_4^{2-} \rightarrow 4\ HCO_3^- + 3Se^0 + 4H_2O + 4H^+$
$CH_4 + SeO_4^{2-} \rightarrow HCO_3^- + Se^0 + H_2O + H^+$
Acetogenesis:
$4\ CH_3OH + 2HCO_3^- \rightarrow 3CH_3COO^- + H^+ + 4H_2O$
$4H_2 + 2HCO_3^- + H^+ \rightarrow CH_3COO^- + 4H_2O$
$CH_3OH + H_2O \rightarrow 3H_2 + CO_2$

1.2. Problem description

Biological treatment of NCG, which is otherwise incinerated during the Kraft pulping process (Prado et al., 2006; Ramirez et al., 2009), is regarded as an efficient and cleaner alternative process for the treatment of waste gases compared to the emissions resulting from the incineration process. Aerobic biofilters (BF) and biotrickling filters (BTF) have been used for the treatment of methanol-rich waste gases (Prado et al., 2004; Prado et al., 2006; Ramirez et al., 2009; Rene et al., 2010). Compared to the aerobic treatment, wherein the final oxidation product of methanol is CO_2, anaerobic treatment offers a platform for the recovery of resources from waste gases through the production of valuable products (Lee et al., 2014). To the best of our knowledge, so far, no studies have reported the use of anaerobic BTFs for the removal of gas-phase methanol. Through anaerobic operation of BTF for gas-phase methanol removal, along with the reduction of the VOC load of the gaseous effluents, the possibility of carbon capture through the production of by-products like VFA and the reduction of atmospheric CO_2 emissions can be explored.

Methanol is a common electron donor for the biological treatment of industrial effluents laden with oxyanions such as sulfate (Liamleam and Annachhatre, 2007), selenate (Takada et al., 2008) and nitrate (Lai et al., 2016). It is important to explore the effect of their elevated concentrations on the biomass used for their bioreduction. Although the effect of sulfate has been thoroughly studied (Hu et al., 2015; Paulo et al., 2004), very little is known on the effects of elevated selenate concentrations on the selenate reducing biomass. Furthermore, the application of methanol-rich waste gases as carbon source for the treatment of effluents laden with selenate and thiosulfate has not been reported before forming the basis of this thesis.

Additionally, numerous studies have reported the anaerobic biological treatment of synthetic methanol-rich wastewaters (Lin et al., 2008; Vallero et al., 2004; Weijma et al., 2003; Weijma and Stams, 2001; Zhou and Nanqi, 2007). However, only few studies have focused on the treatment and resource recovery from a real industrial effluent such as foul condensate (FC) from the Kraft pulping process (Badshah et al., 2012). FC typically contain up to 12 g/L methanol along with ethanol, acetone, terpenes and sulfides (Meyer and Edwards, 2014b), with methanol contributing to about 94-96% of the total chemical oxidation demand (COD) (Liao et al., 2010).

Similar to methanol in the pulp and paper industries, methane is an important constituent of gaseous emissions from petroleum and coal based industries. Additionally, with the worldwide increase in the production of biogas from anaerobic digestion and as a natural by-product of petroleum cracking, methane is an abundant resource and a potential electron donor for the biological removal of nutrients (e.g., sulfate, nitrate and selenate) from synthetic wastewaters (Bhattarai et al., 2017). The utilization of methane for the reduction of sulfate (Bhattarai et al., 2018b), thiosulfate (Cassarini et al., 2017) and nitrate (Lai et al., 2016) has been reported in the literature. However, the application of methane as an electron donor for the reduction of selenate in a BTF was never demonstrated.

VFA are an important by-product formed during the anaerobic treatment of methanol-rich effluents from the pulp and paper industry, and other high organic content waste streams like sewage sludge, food and fermentation waste (Eregowda et al., 2018; Zhou et al., 2017). Recovery of these waste derived VFA through physico-chemical techniques like liquid-liquid extraction (Reyhanitash et al., 2016), electrodialysis (Pan et al., 2018) and adsorption (Rebecchi et al., 2016; Reyhanitash et al., 2017) have been reported in the literature. However, the separation of individual VFA from a VFA mixture is often challenging due to the common functional group (carboxylate). Furthermore, studies on the equilibrium, adsorption kinetics, mechanism and isotherm of the VFA adsorption on anion exchange resins are essential for the selection of suitable anion exchange resins and to design continuous reactors for VFA adsorption and recovery. These are currently lacking in the literature.

In this regard, this PhD research seeks to answer the following research questions:

i) How does the presence of sulfur and Se oxyanions affect methanol utilization for VFA or methane production?

ii) Can gas-phase methanol be treated anaerobically to recovery valuable by-products such as VFA or methane?

iii) Can gas-phase methanol be utilized as electron donor and carbon source for the treatment of S and Se rich industrial effluents?

iv) Can methane be utilized as an electron donor for the reduction of Se oxyanions?

v) Can a methanol-rich effluent such as foul condensate be utilized for VFA production?

vi) What is the strategy to separate individual fatty acids from each other?

1.3. Research objectives

The main objective was resource recovery from methanol rich waste gases and wastewaters. Experiments were performed in anaerobic batch and continuous bioreactor configurations to ascertain the activity of the biomass, methanol/methane removal rates and VFA production. Accordingly, the specific objectives of this research were:

i) BTFs for the anaerobic treatment of gaseous pollutants along with VFA production in the presence of oxyanions of sulfur or selenium.

   a. To investigate the anaerobic removal of gas-phase methanol and reduction of thiosulfate in a BTF along with VFA production.

   b. To investigate selenate reduction and VFA production in an anaerobic BTF fed with gas-phase methanol.

   c. To investigate anaerobic selenate reduction coupled to methane oxidation in a BTF using marine lake sediment as the inoculum.

ii) UASB reactors for the treatment of methanol-rich FC from the Kraft pulping process.

   a. To enrich acetogens in methanol fed UASB reactors operated at 22, 37 and 55 °C.
   b. Acetogenesis of the FC for VFA production.

iii) Examine the effect of selenate, thiosulfate or sulfate on methanol utilization for VFA and methane production, and to compare the anaerobic reduction of selenate and thiosulfate with sulfate in batch systems.

iv) VFA recovery through adsorption on anion exchange resins in batch systems.

   a. To determine the equilibrium kinetics and mechanism of VFA adsorption.
   b. To study VFA adsorption isotherms in single and multi-component systems.
   c. To investigate the selective recovery of acetate from a VFA mixture.

1.4. Thesis structure

This dissertation comprises of nine chapters. ***Figure 1.1*** presents an overview of the structure of the thesis. A brief outline of each chapter is as follows:

Chapter 1 presents the general background, problem statement, research objectives and the thesis structure. **Chapter 2** provides a review of the literature on methanol and methane-rich industrial gaseous and methanol-rich liquid effluents, the current biotechnological approaches for their treatment, presence of sulfur (S) and selenium (Se) oxyanions in industrial effluents, current status of waste derived VFA, the potential benefits of resource recovery through VFA production and the conventional VFA recovery techniques.

Chapter 3 examines the comparative utilization of methanol by activated sludge and the reduction rates in batch systems in the presence of different concentrations of sulfate, thiosulfate and selenate, along with the mass balance. **Chapter 4** investigates the gas-phase methanol removal in a BTF operated continuously for 123 d in order to test the influence of empty bed residence time, intermittent liquid trickling with 6 h and 24 h of wet and dry periods and operation without pH regulation on the elimination capacity and removal efficiency of the BTF in terms of gas-phase methanol removal, $S_2O_3^{2-}$ reduction rate and production of different VFA. In **Chapter 5** gas-phase methanol removal along with selenate reduction in a BTF (inoculated with activated sludge) anoxically operated for 89 d under conditions like step- and continuous-feed of selenate was demonstrated along with the estimation of reduced Se entrapped in the filter bed. **Chapter 6** explores selenate reduction coupled to methane oxidation by anaerobic marine lake sediment in batch enrichments and in a BTF.

For the studies on acetogenesis of FC in **Chapter** 7, the enrichment of acetogenic biomass was carried out in three UASB reactors operated at 22, 37 and 55 °C with methanol-rich synthetic wastewater. The operational parameters that have been reported to inhibit methanogenesis and induce acetogenesis were applied. The enrichment was followed by a test phase wherein the UASB reactors were fed with FC, comprising mainly of methanol, acetone, ethanol dimethylsulfide, methanethiol and α-pinene. The removal of these compounds and the production of VFA were monitored.

Chapter 8 examines VFA recovery through adsorption on anion exchange resins in batch systems. The study involved screening of 12 anion exchange resins, evaluating the adsorption kinetics and adsorption isotherm for single- and multi-component systems. Based on the trend of the adsorption capacity of Amberlite IRA-67 and Dowex optipore L-493, a sequential batch process was tested to separate acetate from a mixture of VFA. **Chapter 9** presents the general discussion and conclusions on the knowledge gained from the above studies along with recommendations and prospects for future research.

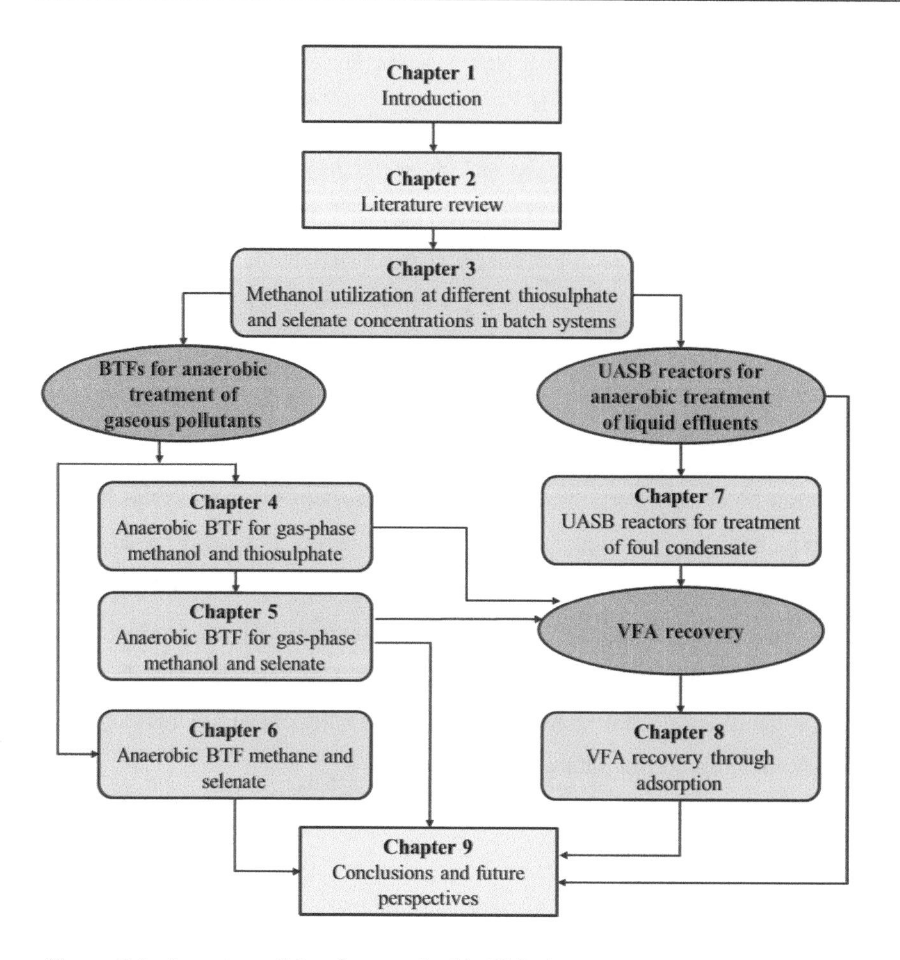

Figure 1.1: *Overview of the chapters in this PhD thesis.*

References

Amon, T., Amon, B., Kryvoruchko, V., Machmüller, A., Hopfner-Sixt, K., Bodiroza, V., Hrbek, R., Friedel, J., Pötsch, E., Wagentristl, H., Schreiner, M., Zollitsch, W., 2007. Methane production through anaerobic digestion of various energy crops grown in sustainable crop rotations. Bioresour. Technol. 98, 3204–3212.

Bhattarai, S., Cassarini, C., Rene, E.R., Zhang, Y., Esposito, G., Lens, P.N.L., 2018. Enrichment of sulfate reducing anaerobic methane oxidizing community dominated by ANME-1 from Ginsburg mud volcano (Gulf of Cadiz) sediment in a biotrickling filter. Bioresour. Technol. 259, 433–441.

Cassarini, C., Rene, E.R., Bhattarai, S., Esposito, G., Lens, P.N.L., 2017. Anaerobic oxidation of methane coupled to thiosulfate reduction in a biotrickling filter. Bioresour. Technol. 240, 214–222.

CEPI, 2017. Key Statistics, European pulp and paper industry.

Choudhari, S.K., Cerrone, F., Woods, T., Joyce, K., O'Flaherty, V., O'Connor, K., Babu, R., 2015. Pervaporation separation of butyric acid from aqueous and anaerobic digestion (AD) solutions using PEBA based composite membranes. J. Ind. Eng. Chem. 23, 163–170.

Elvidge, C.D., Basilian, M.D., Zhizhin, M., Ghosh, T., Baugh, K., Hsu, F.C., 2018. The potential role of natural gas flaring in meeting greenhouse gas mitigation targets. Energy Strategy. Rev. 20, 156-162.

Environmental Protection Agency, 2012. Summary report: global anthropogenic non-CO_2 greenhouse gas emissions: 1990 – 2030. EPA 430-S-12-002: 1-24.

Eregowda, T., Matanhike, L., Rene, E.R., Lens, P.N.L., 2018. Performance of a biotrickling filter for anaerobic utilization of gas-phase methanol coupled to thiosulphate reduction and resource recovery through volatile fatty acids production. Bioresour. Technol. 263, 591–600.

Hu, Y., Jing, Z., Sudo, Y., Niu, Q., Du, J., Wu, J., Li, Y.Y., 2015. Effect of influent COD/SO_4^{2-} ratios on UASB treatment of a synthetic sulfate-containing wastewater. Chemosphere 130, 24–33.

Jones, R.J., Massanet-Nicolau, J., Guwy, A., Premier, G., Dinsdale, R., Reilly, M., 2015. Removal and recovery of inhibitory volatile fatty acids from mixed acid fermentations by conventional electrodialysis. Bioresour. Technol. 189, 279–284.

Jones, R.J., Massanet-Nicolau, J., Mulder, M.J.J., Premier, G., Dinsdale, R., Guwy, A., 2017. Increased biohydrogen yields, volatile fatty acid production and substrate utilisation rates via the electrodialysis of a continually fed sucrose fermenter. Bioresour. Technol. 229, 46–52.

Lai, C.Y., Wen, L.L., Shi, L.D., Zhao, K.K., Wang, Y.Q., Yang, X., Rittmann, B.E., Zhou, C., Tang, Y., Zheng, P., Zhao, H.P., 2016. Selenate and nitrate bioreductions using methane as the electron donor in a membrane biofilm reactor. Environ. Sci. Technol. 50, 10179–10186.

Lee, W.S., Chua, A.S.M., Yeoh, H.K., Ngoh, G.C., 2014. A review of the production and applications of waste-derived volatile fatty acids. Chem. Eng. J. 235, 83–99.

Liamleam, W., Annachhatre, A.P., 2007. Electron donors for biological sulfate reduction. Biotechnol. Adv. 25, 452–463.

Liao, B.Q., Xie, K., Lin, H.J., Bertoldo, D., 2010. Treatment of kraft evaporator condensate using a thermophilic submerged anaerobic membrane bioreactor. Water Sci. Technol. 61, 2177–2183.

Lin, B., 2008. The basics of foul condensate stripping. Krat recovery short course, Florida, USA

Lin, Y., He, Y., Meng, Z., Yang, S., 2008. Anaerobic treatment of wastewater containing methanol in upflow anaerobic sludge bed (UASB) reactor. Front. Environ. Sci. Eng. China 2, 241–246.

Mehdi Y., Hornick J.L., Istasse L., Dufrasne I., 2013. Selenium in the environment, metabolism and involvement in body functions. Molecules. 18, 3292–3311.

Meyer, T., Edwards, E.A., 2014. Anaerobic digestion of pulp and paper mill wastewater and sludge. Water Res. 65, 321–349.

Pan, X.R., Li, W.W., Huang, L., Liu, H.Q., Wang, Y.K., Geng, Y.K., Lam, P.K.-S., Yu, H.Q., 2018. Recovery of high-concentration volatile fatty acids from wastewater using an acidogenesis-electrodialysis integrated system. Bioresour. Technol. 260, 61–67.

Paulo, P.L., Vallero, M.V.G., Treviño, R.H.M., Lettinga, G., Lens, P.N.L., 2004. Thermophilic (55°C) conversion of methanol in methanogenic-UASB reactors: influence of sulphate on methanol degradation and competition. J. Biotechnol. 111, 79–88.

Pol, L.W.H., Lens, P.N.L., Stams, A.J.M., Lettinga, G., 1998. Anaerobic treatment of sulphate-rich wastewaters. Biodegradation 9, 213–224.

Prado, Ó.J., Veiga, M.C., Kennes, C., 2006. Effect of key parameters on the removal of formaldehyde and methanol in gas-phase biotrickling filters. J. Hazard. Mater. 138, 543–548.

Prado, Ó.J., Veiga, M.C., Kennes, C., 2004. Biofiltration of waste gases containing a mixture of formaldehyde and methanol. Appl. Microbiol. Biotechnol. 65, 235-242.

Ramirez, A.A., Jones, J.P., Heitz, M., 2009. Control of methanol vapours in a biotrickling filter: Performance analysis and experimental determination of partition coefficient. Bioresour. Technol. 100, 1573–1581.

Rebecchi, S., Pinelli, D., Bertin, L., Zama, F., Fava, F., Frascari, D., 2016. Volatile fatty acids recovery from the effluent of an acidogenic digestion process fed with grape pomace by adsorption on ion exchange resins. Chem. Eng. J. 306, 629–639.

Rene, E.R., López, M.E., Veiga, M.C., Kennes, C., 2010. Steady- and transient-state operation of a two-stage bioreactor for the treatment of a gaseous mixture of hydrogen sulphide, methanol and α-pinene. J. Chem. Technol. Biotechnol. 85, 336–348.

Reyhanitash, E., Kersten, S.R.A., Schuur, B., 2017. Recovery of volatile fatty acids from fermented wastewater by adsorption. ACS Sustain. Chem. Eng. 5, 9176–9184.

Reyhanitash, E., Zaalberg, B., Kersten, S.R.A., Schuur, B., 2016. Extraction of volatile fatty acids from fermented wastewater. Sep. Purif. Technol. 161, 61–68.

Siddiqui, N.A., Ziauddin, A., 2011. Emission of non-condensable gases from a pulp and paper mill - a case study. J. Ind. Pollut. Control 27, 93–96.

Suhr, M., Klein, G., Kourti, I., Gonzalo, M.R., Santonja, G.G., Roudier, S., Sancho, L.D., 2015. Best available techniques (BAT) - reference document for the production of pulp, paper and board. Eur. Comm. 1–906.

Takada, T., Hirata, M., Kokubu, S., Toorisaka, E., Ozaki, M., Hano, T., 2008. Kinetic study on biological reduction of selenium compounds. Process Biochem. 43, 1304–1307.

Tan, L.C., Nancharaiah, Y.V, van Hullebusch, E.D., Lens, P.N.L., 2016. Selenium: environmental significance, pollution, and biological treatment technologies. Biotechnol. Adv. 34, 886-907.

TRI Program, 2018. 2017 TRI Factsheet for Chemical METHANOL.

Vallero, M.V.G., Camarero, E., Lettinga, G., Lens, P.N.L., 2004. Thermophilic (55-65°C) and extreme thermophilic (70-80°C) sulfate reduction in methanol and formate-fed UASB reactors. Biotechnol. Prog. 20, 1382–1392.

Wang, K., Yin, J., Shen, D., Li, N., 2014. Anaerobic digestion of food waste for volatile fatty acids (VFAs) production with different types of inoculum: Effect of pH. Bioresour. Technol. 161, 395–401.

Weijma, J., Chi, T.M., Hulshoff Pol, L.W., Stams, A.J.M., Lettinga, G., 2003. The effect of sulphate on methanol conversion in mesophilic upflow anaerobic sludge bed reactors. Process Biochem. 38, 1259–1266.

Weijma, J., Stams, A.J.M., 2001. Methanol conversion in high-rate anaerobic reactors. Water Sci. Technol. 44, 7–14.

Yousuf, A., Bonk, F., Bastidas-Oyanedel, J.R., Schmidt, J.E., 2016. Recovery of carboxylic acids produced during dark fermentation of food waste by adsorption on Amberlite IRA-67 and activated carbon. Bioresour. Technol. 217, 137–140.

Zhou, M., Yan, B., Wong, J.W.C., Zhang, Y., 2017. Enhanced volatile fatty acids production from anaerobic fermentation of food waste: A mini-review focusing on acidogenic metabolic pathways. Bioresour. Technol. 248, 68–78.

Zhou, X., Nanqi, R., 2007. Acid resistance of methanogenic bacteria in a two-stage anaerobic process treating high concentration methanol wastewater. Front. Environ. Sci. Eng. China 1, 53–56.

CHAPTER 2

Literature review

2.1. Methanol in the pulping industry

Methanol is a volatile organic compound (VOC), toxic and flammable, naturally produced during the decaying process of organic matter. Process industries such as pulp and paper production, paint manufacturing and petroleum refineries emitted up to 55377 tonnes of methanol in the year 2017 to the atmosphere in the United States alone and at least 65% of the total emission was from Kraft mills of the pulp and paper industries (TRI Program, 2017). During the chemical pulping process (***Figure 2.1***), gas-phase methanol accounts for ~ 80-96% of the emitted chemical oxygen demand (COD) (Tran and Vakkilainnen, 2008; Lin, 2008).

In the Kraft process, white liquor, a mixture of sodium hydroxide (NaOH) and sodium sulphide (Na_2S), is used to separate lignin from the cellulose fibres (compound of interest for paper making). Lignin is a phenolic polymeric structure present in plants that supports vascular plant wood. The pulp washing results in a stream of weak black liquor mainly comprising of recoverable chemicals and lignin. Weak black liquor is sent to the Kraft recovery system, where the inorganic pulping chemicals are recovered for reuse, while the dissolved organics are used as a fuel to make steam and power. In the Kraft process, weak black liquor is concentrated in multi-effect evaporators and concentrators to a point where it can be effectively burned in a recovery boiler (65% solids or higher). The condensates from different stages of condensation and evaporation are collectively known as Kraft condensates (Badshah et al., 2012; Tran and Vakkilainnen, 2012). Methanol is generated in the Kraft process by the removal of a methyl group (de-methylation) from lignin and xylan (hemicellulose groups in the plant cell walls) (Zhu et al., 1999).

Due to the presence of reduced sulphur compounds like methyl mercaptan (CH_3SH), dimethyl sulfide (CH_3SCH_3) and dimethyl disulfide (CH_3SSCH_3), CH_3OH from the Kraft condensates is of poor quality. As a means to save energy and to reduce both the fresh water intake and organic load to the wastewater treatment plant, the condensate is subjected to a stripping process. During the stripping process, methanol and other volatile organic compounds (VOC) from the condensate accumulate as non-condensable gases (NCG) and are generally incinerated (Siddiqui and Ziauddin, 2011; Suhr et al., 2015), leading to the release of malicious exhaust. Thus, other cost-effective and eco-friendly technologies such as biological treatments are interesting for either removal of CH_3OH (Barcón et al., 2012) or resource recovery through the production of biogas (Badshah et al., 2012) or VFA (Eregowda et al., 2018). Bioreactor technologies have been studied for the removal of many VOC and volatile inorganic compounds (VIC) (Iranpour et al., 2005).

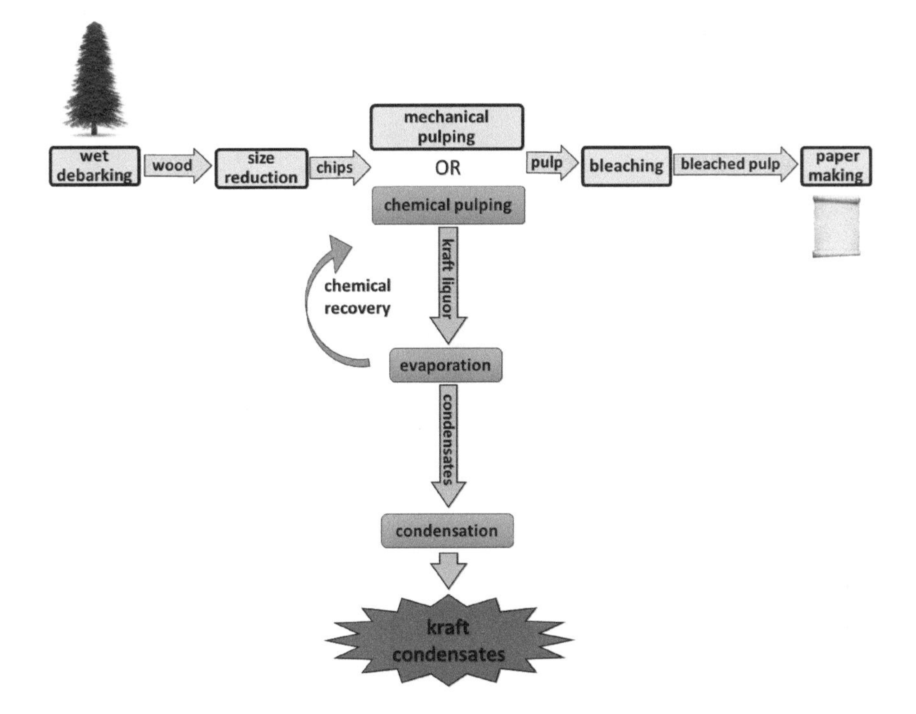

Figure 2.1: *Overview of the chemical pulping process and origin of Kraft condensate in a chemical pulping mill.*

2.2. Bioreactors for gas-phase methanol degradation

Biological processes are effective for the treatment of odorous gases and VOC from different manufacturing industries (Mudliar et al., 2010). The common types of biological techniques for gas-phase methanol treatment are bioscrubbers, biofilters and biotrickling filters (BTF) ***(Figure 2.2)***. These three reactor configurations are based on the aerobic biodegradation of methanol ($\frac{1}{6}\ CH_3OH + \frac{1}{4}O_2 \rightarrow \frac{1}{6}CO_2 + \frac{1}{3}H_2O$) and they have a nearly similar pollutant removal mechanism, except for the mode of attachment of the microorganisms (either suspended in the liquid-phase or immobilized on a carrier), packing media and pollutant concentrations.

During the aerobic utilization, methanol is mineralized to carbon dioxide and water by aerobic methylotrophic families, such as *Pseudomonas, Methylomonas, Bacillus, Protaminobacter,* and *Arthrobacter.* These microorganisms use methanol as the electron donor and carbon source and oxygen as electron acceptor. Performance of these bioreactors is influenced by parameters such as the moisture content, structure of the packing material, nutrient composition, pH control, pressure drop and transient/shock loads (Kim et al., 2008; Kim & Deshusses, 2005; Mudliar et al., 2010).

2.2.1. Biofilters

Biofilters (***Figure 2.2a***) are used to remove odorous pollutants from sewage treatment facilities, composting and other industrial wastewater treatment plants. The basic working principle is similar to that of a biofilm process wherein gas-phase pollutants are absorbed by a liquid-phase and diffuse to the biofilm interphase (Rene et al., 2012). The porous packed medium of the bioreactor supports the active microbial community, which converts the pollutant into carbon dioxide, water, and other intermediate products. Biofilters offer advantages such as better nutrient distribution, high porosity, high moisture retention capacity and buffering capacity and the disadvantages include clogging and difficulty in controlling pH and moisture (Kim et al., 2008; Mudliar et al., 2010). Depending on the type of packing material, empty bed residence time (EBRT), specific surface area and other operational parameters, a methanol removal efficiency of up to 97% can be achieved using biofilters.

2.2.2. Bioscrubbers

A bioscrubbing unit (***Figure 2.2b***) involves the absorption of pollutants by the liquid in a scrubber followed by biodegradation in a bioreactor (Cabrera et al., 2011; Mudliar et al., 2010). The system contains a re-circulating liquid and suspended activated sludge that is sufficiently aerated for the complete oxidation of gaseous pollutants. The effluent from the bioreactor is re-circulated to the absorption column. The addition of inert packing material in the absorption column enhances the transfer of gas into the liquid-phase.

2.2.3. Biotrickling filters

A biotrickling filter (BTF) is a fixed bed reactor wherein the pollutants are passed through a microbial consortium immobilized on a suitable support material with high surface area (***Figure 2.2c***). The recirculation of water in the system maintains the required humidity of the packing bed and delivers nutrients for the microbes. The trickling liquid also acts as a pH buffer for the trickling filter medium (Kim et al., 2008). Gaseous pollutants are absorbed by the aqueous phase, which further diffuse to the biofilm where they are subsequently degraded by the microbes (Massoudinejad et al., 2008; Groenestijn & Kraakman, 2005).

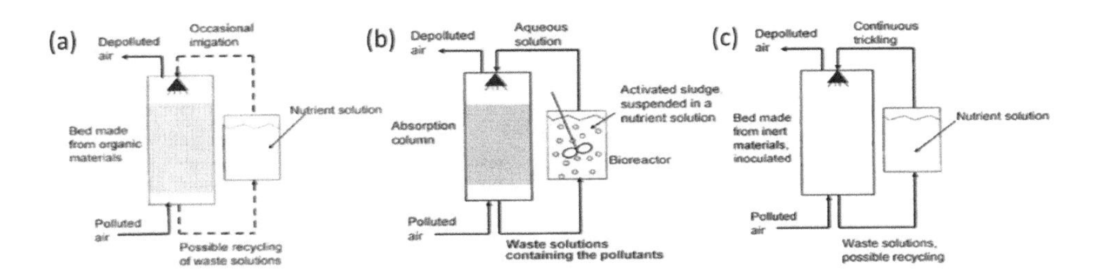

Figure 2.2: *Schematic comparison of (a) biofilter, (b) bioscrubber and (c) biotrickling filter (Mudliar et al., 2010).*

The efficiency of BTFs depends on the operational parameters such as the type of packing material, pH, temperature, inlet gas loading rate, type of biomass and nutrient loading rate (Fernández et al., 2013). Typical packing materials for a BTF can be made of ceramic or plastic structures, activated carbon, lava rock, perlite, peat, wood chips, cellite, pall rings, polyurethane foam or mixtures of inert materials (Rene et al., 2010a; Kennes and Thalasso, 1998).

Table 2.1: *Maximum elimination capacity (EC_{max}) of gas-phase methanol at different inlet loading rates (ILR) and empty bed residence time (EBRT) in different reactor configurations.*

Gas-phase composition	**Reactor type and operation**	**Methanol ILR ($g/m^3.h$)**	**EBRT (s)**	**Methanol EC_{max} ($g/m^3.h$)**	**References**
Methanol	Anaerobic BTF, thiosulfate rich trickling liquid	25-30	138	21	Chapter 4
Methanol	Anaerobic BTF, selenate rich trickling liquid	7-56	276	46	Chapter 5
Methanol + formaldehyde	Aerobic BF and BTF	Up to 644.1	20.7-71.9	600	Prado et al. (2004)
Methanol + H_2S	Aerobic single stage BTF, low pH	240-260	24	236	Jin et al. (2007)
Methanol + H_2S + α-pinene	Aerobic two stage: BF and BTF	28-1260	41.7	894	Rene et al. (2010)
Methanol	Aerobic BTF	3700	20-65	2160	Ramirez et al. (2009)
Methanol + formaldehyde	Aerobic BTF	Up to 700	36	552	Prado et al. (2006)

Several studies (***Table 2.1***) have been conducted on ascertaining the effectiveness of aerobic BTFs for the removal of volatile organic compounds (VOC). Stoichiometric amounts of oxygen are required for the microbial oxidation of methanol (or other VOC) to CO_2. Mixtures of various organic and inorganic compounds have also been tested in BTFs (Lopez et al., 2013; Philip & Deshusses, 2008; Tsang et al., 2008; van Leerdam et al., 2008). Several studies have reported the aerobic treatment of methanol in BTFs at different inlet loading rate (ILR) and empty bed residence time (EBRT). Anaerobic BTFs have been mainly tested for H_2S removal and removal efficiencies in the range of 80-99% have been reported (Rattanapan et al., 2012; Syed et al., 2009). An anaerobic BTF was operated by Eregowda et al. (2018) for the gas-phase methanol treatment, along with the reduction of thiosulfate rich synthetic wastewater as the trickling medium. The gas-phase methanol was converted to acetate and accumulated in the tricking liquid (Eregowda et al., 2018).

2.2.4. Operational parameters for the biotrickling filter (BTF)

The physicochemical parameters influencing the efficiency of anaerobic BTF operation are described in the following sections.

2.2.4.1. Packing material

Appropriate packing materials should offer a large surface area, a high water retention capacity without becoming saturated, low bulk density, high porosity and structural integrity (Rene et al., 2012; Kim & Deshusses, 2008). Physical characteristics of common packing materials are listed in ***Table 2.2***. Several types of synthetic packing materials are effective in the removal of waste gases. For example, a maximum hydrogen sulphide gas removal efficiency of 99% at a loading rate of 120 g S m^3/h was achieved using polypropylene pall rings (Fernández et al., 2013). A mixture of 50% polyurethane foam cube and 50% plastic pall rings have been used in methanol or methane fed anaerobic BTFs (Cassarini et al., 2017; Eregowda et al., 2018).

Table 2.2: *Characteristics of materials commonly used for filter bed packing (Kim and Deshusses, 2008).*

Property	Shape	Dimension (mm)	Surface area (m^2/m^3)	Porosity (%)	Bulk density kg/m^3
Lava rock	Irregular	1-40	-	49	-
Pall ring	Cylinder	25	188	90	141
Polyurethane foam cube	Cube	40	600	98	28
Porous ceramic beads	Beads	4	2500	38	-
Porous ceramic raschig ring	Raschig rings	13	360	54	-
Compost-wood chips mixture	Irregular	-	-	40-60	-

2.2.4.2. Moisture content, pH and temperature

The filter bed moisture content is one of the basic parameters that affects the long and short-term performance of BTFs. Microorganisms in the BTF system require sufficient moisture for their metabolism. Depending on the type of packing material used and biomass applied, the moisture content of the filter bed ranges between 35% to 80% (Dorado et al., 2010; Gribbins and Loehr, 1998). Although the waste gas that enters the BTF is saturated with water, it becomes unsaturated when the temperature rises. Hence, it is important to maintain the optimal humidity levels within the BTF, and also to compensate for the moisture losses. Conversely, too much moisture leads to a slow mass transfer of gaseous compounds into the biofilm and increased pressure drop across the packed bed (Rattanapan and Ounsaneha, 2012).

A change in pH can strongly affect the metabolic activity of the microorganisms. Each species is most active over a certain pH range and will be inhibited if the conditions deviate too much from their optimal range. For an anaerobic process, higher sulphate and thiosulphate reducing activity is achieved within the pH range of 7.0 to 8.0 (Hu et al., 2015).

The microbial growth rate in a BTF is a strong function of the filter bed temperature. The temperature determines the type of the microbial community that prevails within the system. Temperature in the BTF is mainly influenced by the inlet air temperature and the temperature produced by the biological reaction within the trickling bed (Chung et al., 1998). Most biotrickling processes operate at a temperature range of 20 to 45 °C, with 35 to 37 °C often noted as the optimal range (Wani et al., 1999). Some studies have worked with thermophilic BTFs operating in the temperature range of 45 to 75 °C (Dhamwichukorn et al., 2001). In case of anaerobic methanol degradation, thermophilic biomass thrive at up to 55 °C using sulphate or thiosulphate as an electron acceptor (Paulo et al., 2004).

2.2.4.3. Empty bed residence time (EBRT)

EBRT (min) is the ratio of the volume of empty trickling bed to the flow rate of the gas-phase (Devinny et al., 1999). The EBRT determines the removal efficiency of the pollutant in a BTF. A higher EBRT allows the pollutants to stay longer in the BTF for microbial degradation. Studies on H_2S treatment recorded a maximum elimination capacity at an EBRT of ~ 30 seconds (Montebello et al., 2012). In anaerobic BTFs, data regarding the effect of EBRT on gas-phase pollutant (such as methanol) removal is unavailable.

2.2.4.4. Inlet loading rate (ILR)

ILR ($g/m^3.h$) is the ratio of the concentration of a certain pollutant, such as methanol, to the EBRT of the BTF. The BTF operation is strongly controlled by the mass transfer rate of the gaseous pollutants and their associated biodegradation rates. At lower pollutant loads, the biodegradation rate is larger than the mass transfer rate and the mass transfer rate becomes larger than the degradation rate at higher loading rates. Hence, the pollutant degradation rate depends on the mass transfer rate of the pollutants through the trickling media. The ILR depends on both the flow rate and pollutant concentration (Li et al., 2014).

2.2.4.5. Shock loading

A shock loading ($g/m^3.h$) is the loading of excess concentrations of waste gases into a BTF. The quantity of pollutant in the BTF can affect the pollutant uptake rate by microbes. The availability of high inlet concentrations or increased gas flow rates creates a shock loading effect onto the biomass. On the contrary, too low pollutant loading rates in the BTF system create starvation conditions. Both shock loading and starvation affect the biomass activity in the system (Chen et al., 2014).

2.3. Anaerobic methanol utilization

2.3.1. Methanogenesis

Anaerobic methylotrophy has been studied extensively with a focus on enhancing methanogenesis in upflow anaerobic sludge blanket reactor (UASB) reactors in the liquid-phase (Lin et al., 2008; Paulo et al., 2004; Weijma et al., 2003) for the production of biogas. Anaerobic mineralization of methanol occurs via direct conversion to methane by methylotrophic methanogens, indirect conversion to acetate by acetogens coupled to acetoclastic methanogenesis or indirect conversion to H_2 and CO_2 coupled to hydrogenotrophic methanogenesis (Fermoso et al., 2008). Methanogenesis is limited by high organic loading rates of methanol, low pH or acetate accumulation (Weijma and Stams, 2001). As VFA concentrations increase, the reactor becomes unstable and hence affects the removal efficiency (%) and elimination capacity ($g/m^3.h$) of the reactor. According to Florencio et al. (1997), as VFA concentrations increase, the reactor becomes unstable, hence affect the removal efficiency and elimination capacity of the reactor. The increase in the VFA concentration in the reactor could be an economic benefit if these VFA are extracted before reducing the removal efficiency and elimination capacity of the reactor.

While most of the studies on methanol mineralization are carried out using synthetic wastewater, few studies have demonstrated the treatment of methanol-rich effluents such as Kraft condensates (Badshah et al., 2012; Dias et al., 2005; Dufresne et al., 2001). For example, a methanol condensate fed UASB reactor operated for 480 d showed a COD removal efficiency of 84-98% at an OLR of 5-11 gCOD/L with a biogas production of 0.29 L CH_4/gCOD (Badshah et al. 2012). However, no study on the VFA production from the methanol rich industrial effluents has been so far reported.

2.3.2. Methanol as electron donor for the removal of sulfur oxyanions

Along with SO_4^{2-}, pulp and paper wastewaters contain $S_2O_3^{2-}$ (Aylmore and Muir, 2001) since $S_2O_3^{2-}$ is used to remove excess chlorine in the paper and textile industry. $S_2O_3^{2-}$ is also a key intermediate during SO_4^{2-} reduction. The metabolic pathways of $S_2O_3^{2-}$ reduction have been rarely studied. The sulfate reducing bacteria (SRB) that are capable of utilizing methanol for SO_4^{2-} reduction are also capable of $S_2O_3^{2-}$ reduction. Thus, methanol-rich effluents such as NCG or Kraft condensates could be potential substrates for the treatment of S oxyanion rich wastewaters.

2.3.3. Methanol as electron donor for the removal of selenium oxyanions

Selenium is a trace element that is naturally found in the earth's crust. The toxicity of Se is mainly due to the generation of free radical species that induce DNA damage and Se reactivity with thiols that affect the function of DNA repair proteins. Furthermore, the substitution of seleno-aminoacids in the place of its sulfur analogues also influences the intercellular enzyme activity (Lenz and Lens, 2009).

About 40% of the selenium emissions to atmospheric and aquatic environments are caused by mining-related operations (Tan et al., 2016). Several studies demonstrated aerobic and anaerobic bioreduction and recovery of Se oxyanions in UASB reactors (Dessì et al., 2016), membrane bioreactors (MBR) (Lai et al., 2016), CSTR (Staicu et al., 2017) and batch systems (Espinosa-Ortiz et al., 2015; Mal et al., 2017).

The mechanisms of dissimilatory reduction of SeO_x using electron donors like lactate, acetate, glucose, pyruvate and H_2 have been elucidated (Nancharaiah and Lens, 2015). In the absence of the dissimilatory reduction mechanism, the microbes reduce SeO_x to Se^0 as a detoxification strategy wherein the toxic, soluble SeO_4^{2-} is reduced to Se^0, facilitated by biomolecules like glutathione, glutaredoxin, and siderophores. Furthermore, SeO_3^{2-} reduction is also catalysed by other terminal reductases like nitrite reductase or sulfite reductase (Khoei et al., 2017; Nancharaiah and Lens, 2015; Tan et al., 2016a). Use of methanol as an electron donor for bioreduction of S and Se oxyanions and nitrates is reported at lab scale (Fernández-Nava et al., 2010; Soda et al., 2011; Takada et al., 2008). Thus, for the industrial effluents from selenium, gold and lead mining industries that contain up 620, 33 and 7 mg/L SeO_x, respectively (Pettine et al., 2015), methanol is a suitable electron donor.

2.4. Waste derived volatile fatty acids

VFA are hydrocarbons with a carboxylic functional group (COO^-) that find their application in a wide range of industries like food, fodder, disinfection, tanning and pharmaceutical industries and as precursors for biopolymers, biofuels and other chemicals such as esters, ketones, aldehydes and alkanes (Da Silva and Miranda, 2013; Eregowda et al., 2018; Rebecchi et al., 2016). Currently, the major fraction of the global VFA demand is met through petrochemical routes (Reyhanitash et al., 2017). Considering the fact that VFA, especially acetate, butyrate and propionate are common by-products and intermediates in the anaerobic digestion and fermentation of organic material, biogenic VFA production holds a huge potential, if teamed with an efficient recovery process.

2.4.1. VFA recovery techniques

Physicochemical downstream processing techniques such as crystallization, adsorption, esterification, distillation, dialysis, liquid-liquid extraction have been applied for the recovery of VFA (Jones et al., 2015; Martinez et al., 2015; Pan et al., 2018; Reyhanitash et al., 2016; Yousuf et al., 2016; Zhang and Angelidaki, 2015).

2.4.1.1. Electrodialysis

In the electrodialysis process, the negatively charged VFA migrate towards the positively charged electrodes when an electrical voltage is applied between the electrodes. The anion exchange membrane (AEM) allows only the negatively charged VFA to pass, while retaining the cations. Thus, other anions and VFA are concentrated in the product chamber. Electrodialysis has been used to recover VFA efficiently (up to 99%) from various fermentation broths and sludge digestates (Jones et al., 2015; Pan et al., 2018).

2.4.1.2. Solvent extraction

Liquid-liquid (or solvent) extraction is a common downstream process in the chemical industries. Solvents such as alcohols, ketones, ethers, aliphatic hydrocarbons, organophosphates such as trioctyl-phosphine oxide (TOPO), tri-n-butyl phosphate and aliphatic amines are commonly used for the phase separation of VFA or other ionic compounds. Parameters such as the solubility of the VFA extracted and the concentration of the extractant influence the efficiency of organic acid extraction (Alkaya et al., 2009; Yang et al., 1991).

Furthermore, supercritical fluids and ionic liquids (ILs) are considered as promising alternative solvents for green separation processes. Because of the properties like adjustable solubility with negligible vapour pressures, relatively high viscosity, high thermal and electrochemical stability and ionic liquids (i.e., salts with melting points close to room temperature) are emerging as green alternatives to volatile organic solvents (Djas and Henczka, 2018; Liu et al., 2016). Parameters like the molecular structure of the carboxylic acids, concentration, type and concentration of the complexing reactant (such as a mixture of tri-n-octyl/n-decylamine, tri-n-hexylamine, and a secondary long chain amine LA-2), pH of the aqueous phase, and temperature play a key role in the extraction of VFA using supercritical fluids and ionic liquids as solvents (Djas and Henczka, 2018).

2.4.1.3. Adsorption

Several studies report the recovery of VFA such as acetic, propionic and butyric acids using a variety of adsorbent materials ranging from activated carbon, zeolites and anion exchange resins (AiER) (Da Silva and Miranda, 2013; Li et al., 2015; López-garzón and Straathof, 2014; Pinelli et al., 2016; Reyhanitash et al., 2017; Yousuf et al., 2016). Adsorption on AiER is an easy strategy for VFA recovery due to the relative simplicity of the technique, low cost and ease of operation and scale-up (Pinelli et al., 2016).

2.4.2. VFA recovery by adsorption on ion exchange resins

2.4.2.1. VFA adsorption on AiER

Since adsorption is a surface phenomenon, the surface area of the adsorbent material needs to be of high magnitude. For example, the surface area of Dowex optipore L 493, a commercial polymeric resin, is 1100 m^2/g (Nielsen et al., 2010). Adsorption of VFA on surface active materials such as silica gel, zeolites and activated carbon is based on the physical, hydrophobic interaction and the Van der Waals forces between the adsorbent and adsorbate materials

(Levario et al., 2012). Adsorption on ion exchange resins, in particular, AiER for VFA is based on the formation of ionic bonds between the ionized functional groups of carboxylic acids and the positively charged functional groups such as ammonium salts on the resin matrix (Rebecchi et al., 2016).

Considering the fact that the adsorption capacity of the resins usually decreases when the VFA mixture is prepared in synthetic wastewater, incorporation of another resin to remove these inorganic anions prior to the VFA adsorbing resins could be beneficial to maintain high adsorption capacities for the VFA. Incorporation of a sequential system with more than one kind of resin can be a promising approach to achieve individual recovery of the ionic compounds, both organic and inorganic (Chapter 8).

2.4.2.2. Adsorption isotherms and modelling

For efficient designing and operation of VFA adsorption on AiER, equilibrium adsorption data, adsorption capacity and adsorption kinetic data are required to predict the performance of the adsorption under varying operating conditions (Da Silva and Miranda, 2013). A variety of adsorption models that predict the effect of the influencing parameters are available and the VFA adsorption behaviour can be predicted by fitting the experimental data to the models.

The capacity of the ion exchange resins to adsorb the VFA can be quantified by fitting the equilibrium experimental data to isotherm models like Langmuir, Freundlich and Brunauer–Emmett–Teller (BET). Important parameters representing the affinity between the adsorbate and adsorbent as well as maximum adsorption capacities can be calculated to gain insight into the adsorption isotherm for single and multi-component systems (Hamdaoui and Naffrechoux, 2007). The primary assumptions for the Langmuir adsorption isotherm model are: (1) adsorption is monolayer, (2) the adsorption sites are finite, localised, identical and equivalent and (3) no lateral interaction or steric hindrance occurs between the adsorbed molecules, nor between the adjacent sites (Foo and Hameed, 2010; Guerra et al., 2014).

The Freundlich isotherm describes the non-ideal, reversible adsorption, and can be applied for multilayer adsorption over heterogeneous surfaces, wherein the amount adsorbed is the summation of the adsorption on all the sites. The stronger sites are occupied first until the adsorption energy is exponentially decreased until the completion of the adsorption process (Ahmaruzzaman, 2008; Nancharaiah, 2010). The BET isotherm was developed to derive multilayer adsorption systems and is widely applied in gas-solid equilibrium systems (Zhang et al., 2014). The model assumes that for each adsorption site, an arbitrary number of molecules may be accommodated (Passe-Coutrin et al., 2008). The BET model can be extended to liquid-phase adsorption by replacing the partial pressure of the gas-phase adsorbate with the equilibrium concentration of the liquid-phase adsorbate.

2.4.3. Parameters influencing VFA adsorption on ion exchange resins

2.4.3.1. Resin selection

Among the wide range of adsorbents proposed for VFA adsorption, type 2 (weakly basic) anion exchange resins are the most suitable for VFA adsorption because the functional group (tertiary amine) can adsorb carboxylic acids as charge-neutral units to maintain the charge neutrality (Reyhanitash et al., 2017).

2.4.3.2. pH and temperature

Since the ionization of the carboxylic acid is crucial in achieving maximum adsorption during the chemisorption process, the pH of the liquid-phase should necessarily be close to the pKa of the acids. The adsorption of acids by amine groups is exothermic and decreases with an increase in temperature due to a decrease in the adsorption affinity towards adsorbate molecules and a shift in VFA dissociation equilibrium towards the liquid-phase. Thus, the exothermic nature of the reaction promotes desorption with increasing temperature (Pradhan et al., 2017).

2.4.3.3. Presence of other anions

The VFA rich broth will usually contain common anions like nitrates, nitrites, sulphates, phosphates and chlorides. Since the adsorbate-adsorbent affinity depends on the charge density of the adsorbate molecules, these ions will compete with the VFA and affect the VFA recovery (Tan et al., 2018; Reyhanitash et al., 2017).

2.4.3.4. Incubation time

The rates of the VFA adsorption and the rate-limiting steps can be analysed by studying the kinetic models like the pseudo-first, second order or Elovich kinetic model. It is important to study diffusion models like the intraparticle diffusion and the liquid diffusion models to understand the diffusion mechanism of the adsorbate from the bulk liquid to the resin surface since the synthetic ion-exchange resins are usually macroporous polymeric materials with a very large surface area (e.g. surface area of Dowex-optipore L-493 is ~ 1100 m^2/g) and the functional groups are located in the pores and on the surface of the resin. The transport of the adsorbate molecules can be limited by the liquid and/or intraparticle diffusion (Sheha and El-Shazly, 2010).

2.5. Other industrial waste gases

2.5.1. Methane availability and emissions

Though methane deposits are valuable energy resources, their potency as greenhouse gas towards global warming is very high. Globally, over 60% of the total methane emissions come from anthropogenic activities such as petroleum, natural gas and coal industry, agriculture, and waste management activities (Bousquet et al., 2006). It is estimated that 143 billion m^3 of methane gas, 3.5% of the global gas production, is flared annually because of technical, regulatory, or economic constraints (Elvidge et al., 2016). With a lifetime of 12 years in the atmosphere and global warming potential of 28, the comparative impact of methane (trapping of radiation) is 28 times larger than that of CO_2 over a period of 100 years (IPCC, 2014). The globally averaged monthly mean atmospheric methane concentration is increasing at an alarming rate, reaching a concentration of 1850.5 ppb in July 2018 as shown in ***Figure 2.3*** (Dlugokencky, 2018).

Furthermore, biological methane production through anaerobic methanogenesis from organic substrates such as municipal waste, wastewater sludge, landfill leachate, wastewater, agricultural biomass, sludge form dairy and food processing industries are also studied globally both in terms of methane production and waste management. A variety of feedstocks have been used for the anaerobic digestion (AD) at mesophilic conditions (35–40 °C). For instance, one tonne dry feedstock can yield 13 to 635 m^3 of methane gas, depending on the composition of the feedstock (Brown et al., 2012). Municipal wastes usually yield more methane than lignocellulosic biomass such as yard trimmings (Brown et al., 2012). A study evaluating the energy crops for biogas production in the EU-25 shows that 320 million tonnes of crude oil equivalents (COE) could be produced with crop rotations that integrate the production of food, feed and raw materials (e.g. oils, fats and organic acids), which would provide up to 96% of the total energy demand of cars and trucks in the EU-25 (Yang et al., 2014). Owing to its easy availability, methane is a good candidate as energy source for biological processes.

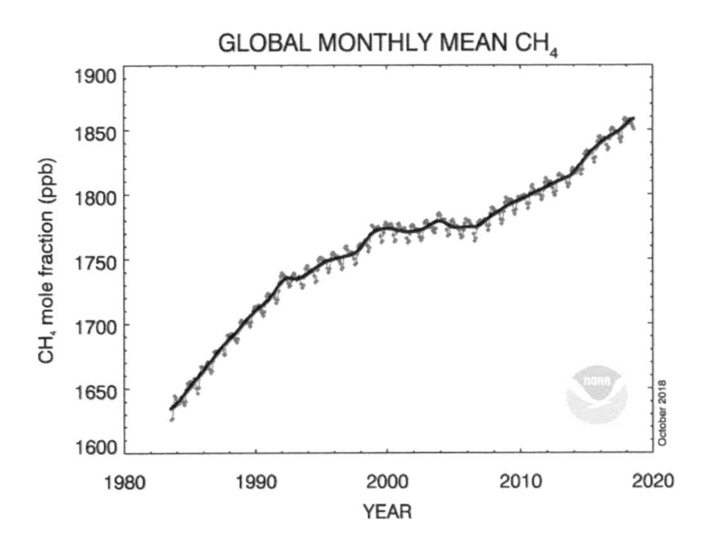

Figure 2.3: *Globally-averaged, monthly mean atmospheric* methane *concentration (Dlugokencky, 2018).*

2.5.2. Aerobic and anaerobic methanotrophy

2.5.2.1. Aerobic methane oxidation

Aerobic methanotrophs or methane oxidizing bacteria use methane as the sole carbon source. To generate energy and assimilate carbon, methane is converted to methanol, catalyzed by particulate methane monooxygenase (pMMO), followed by conversion of methanol to formaldehyde. Formaldehyde is then assimilated into cell biomass (Kang and Lee, 2016; Nazaries et al., 2013).

2.5.2.2. Anaerobic methane oxidation

Anaerobic oxidation of methane (AOM) is a natural phenomenon occurring in deep marine environments with methane seepage. AOM is mediated by a consortium of archaea called anaerobic methanotrophs (ANME) and sulphate reducing bacteria (SRB) (Cassarini et al., 2017). As methane diffuses upwards to shallower sulfate-penetrated sediments, more than 90% of the methane is consumed by ANME, corresponding to around 7-25% of the total global methane production (Bousquet et al., 2006; Mueller et al., 2015). Anaerobic oxidation of methane (AOM) facilitated by ANME can capture up to 300 million metric tonnes of methane per year and plays a key role in regulating the methane global flux and carbon cycle (Soo et al., 2016).

From the perspective of greenhouse gas mitigation, ANME exhibits an extremely useful feature. Their aerobic counterpart incorporate up to half of the metabolized methane carbon into their own biomass, while converting a short-lived, but powerful, greenhouse gas (CH_4) into a weaker but much longer-lived one (CO_2) that also acidifies ocean waters. In contrast, ANME convert methane into carbonate, which reacts with calcium in seawater and precipitates as solid calcium carbonate rock on the sea floor. Where methane venting and anaerobic methanotrophy occur, large plates of microbially derived calcium carbonate beds occur, thereby, permanently sequestering methane carbon (Nazaries et al., 2013; Stolaroff et al., 2012).

Due to their extremely low growth rates with a doubling time of 2-7 months, the biochemistry of methane oxidation and assimilation of methane into anaerobic methanotrophs is poorly understood and the knowledge on regulating factors is limited (Zhang et al., 2010). Furthermore, the poor solubility of methane in seawater (0.055 mg/m^3) (Duan and Mao, 2006; Serra et al., 2006) makes the bioavailability of methane a limiting factor.

2.5.3. Gas-to-liquid fuel technologies

Methane is an easily available, low-priced energy source for nutrient (sulfur, selenium and nitrates) removal in a biological process. Furthermore, conversion of methane to valuable liquid chemicals like methanol, acetate and other next-generation fuels is an effective method of carbon capture to utilize the impure biogas from anaerobic digestion and waste gas from coal and petroleum industries, which is otherwise flared (Fei et al., 2014; Kang and Lee, 2016).

Thermochemical conversion, such as the Fischer–Tropsch process, involves a series of catalytic processes for the conversion of synthesis gas (produced by steam reforming that converts methane into a mixture of carbon monoxide and hydrogen) into a mixture of products that can be refined to synthetic fuels, lubricants and petrochemicals (Yang et al., 2014). Biological conversion of methane is economically and environmentally sustainable, since it requires a smaller foot-print, does not involve cleaning and is technologically less complex (Soo et al., 2016). Compared to thermochemical processes, biological conversion is highly attractive due to efficient conversion reactions under mild operating conditions and the fact that it is least polluting. Furthermore, microorganisms have an extraordinary biological diversity which enables them to adapt to various environments within the toxicity limits (Ge et al., 2014; Taher and Chandran, 2013). ***Figure 2.4*** shows a comparative overview of thermochemical and biological pathways for the conversion of methane to liquid chemicals.

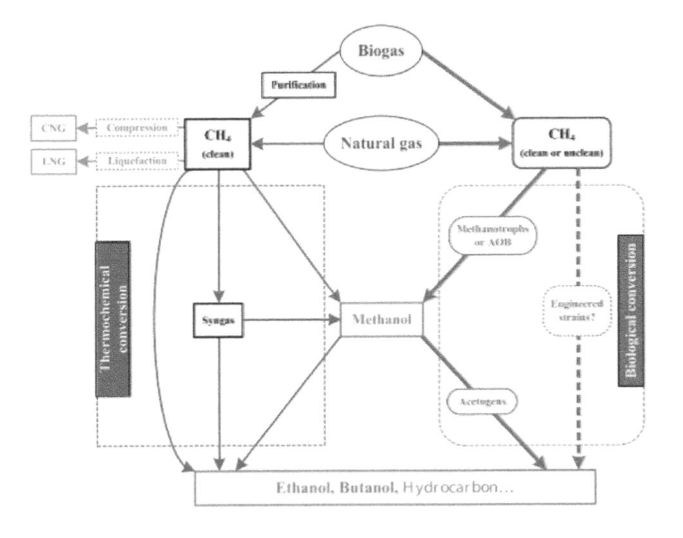

Figure 2.4: *Comparison between the thermochemical and biological conversion pathways for liquid fuel production from methane (Ge et al., 2014).*

2.5.3.1 Indirect bioconversion of methane to liquid fuels

Methane-to-methanol conversion is attracting considerable attention owing to the abundant reserves of methane and the potential for methane utilization as an alternative chemical feedstock for the production of value-added chemicals. Biological pathways may be a good choice for methane-to-methanol conversions because bioconversion can be accomplished in an energy-efficient and environmentally benign manner (Hwang et al., 2014).

The bioconversion of methane to methanol is associated with low energy consumption, high conversion, high selectivity, and low capital costs compared with chemical methods since it is carried out by the methane monooxygenase (MMO) enzyme under mild conditions (Conrado and Gonzalez, 2014). Methanotrophs use MMO to convert methane to methanol, which is further converted to formaldehyde by the enzyme methanol dehydrogenase (MDH) and further to cell biomass through the ribulose monophosphate (RuMP) cycle or oxidised to formate and CO_2 for biosynthesis. Thus, for successful production and accumulation of methanol, suitable inhibitors for MDH are required. Specific and non-specific inhibitors like cyclopropanol, EDTA and high concentrations of sodium chloride and phosphates can be applied for the inhibition of MDH and other enzymes or co-factors in the electron transport chain (Hur et al., 2017; Kim et al., 2010).

Aerobic utilization of methane for methanol production has been demonstrated at lab scale. For example, methanotrophs that are comparable to obligate methanotrophs from the genus *Methylocaldum* were isolated from solid-state anaerobic digestate and were grown on biogas for ethanol production. Several MDH inhibitors and formate as electron donor were used for the synthesis of up to 430 mg/L of methanol in 48 h with a conversion efficiency of 25.5% (Sheets et al., 2016).

2.5.3.2. Direct bioconversion of methane to liquid chemicals

Direct bacterial assimilation of methane for the production of industrial platform chemicals like lactate, acetate and methanol have been reported using genetically engineered methanotrophic organisms (Henard et al., 2016; Sheets et al., 2016; Soo et al., 2016).

Biocatalysis of methane to lactate in a continuously stirred tank bioreactor (CSTR) was demonstrated by heterologous overexpression of *Lactobacillus helveticus* L-lactate dehydrogenase in the methanotrophic bacterium *Methylomicrobium buryatense.* A yield of up to 0.05 g lactate/g methane was reported with a productivity of 8 mg lactate/L.h (Henard et al., 2016). In another study, pure cultures of the engineered archaeal methanogen *Methanosarcina acetivorans* were grown anaerobically on methane for the production of acetate. The enzyme methyl-coenzyme M reductase (Mcr) from ANME-1 was cloned into *M. acetivorans* to effectively run methanogenesis in reverse for production of up to 52 μmol of acetate from 143 μmol of methane in a period of 5 d (Soo et al., 2016).

2.5.3.3. Methane as the electron donor for nutrient removal

A few recent studies have explored the potential of methane as a carbon source and electron donor for the anaerobic reduction of oxyanions like sulphate, thiosulfate, selenate and nitrate. Thiosulfate reduction was demonstrated in a methane fed anaerobic BTF by Cassarini et al. (2017). During the 213 d of BTF operation, a thiosulfate reduction rate of up to 0.38 mM/L.d was achieved (Cassarini et al., 2017). Simultaneous bioreduction of selenate and nitrate was studied in a methane fed MBR (Lai et al., 2016). SeO_4^{2-} bioreduction in a methane fed MBR was carried out by a single methanotrophic genus, such as *Methylomonas,* that performed methane oxidation directly coupled to SeO_4^{2-} reduction, and methanotrophic methane oxidation to form organic metabolites that were electron donors for a synergistic SeO_4^{2-} reducing bacterium.

Denitrifying anaerobic methane oxidation (DAMO) is regarded as a sustainable option for wastewater treatment plants as DAMO does not require expensive electron donors such as acetate, methanol, or ethanol for the denitrification process (Wang et al., 2017). The DAMO archaea convert nitrate to nitrite using electrons derived from methane, while the DAMO bacteria reduce nitrite to nitric oxide and subsequently nitric oxide to nitrogen and oxygen gas via the inter-aerobic denitrification pathway (Wang et al., 2017). Production of C-4 carboxylic acids, specifically crotonic and butyric acids, from methane was illustrated in engineered *Methylomicrobium buryatense* through diversion of the carbon flux through the acetyl-CoA node of sugar linked metabolic pathways using reverse β-oxidation pathway genes (Garg et al., 2018).

Full-scale application of methane driven biological gas-to-liquid fuel technologies and nutrient removal in WWTPs face several challenges like prolonged enrichment period for the methanotrophic biomass, improving biological activity and strengthening process robustness (He et al., 2015). In this direction, enrichment of ANME coupled to sulfate reduction has been successfully demonstrated in MBR and BTF (Bhattarai et al., 2018b; Meulepas et al., 2009). Bhattarai et al. (2018) enriched to more than 50% of the ANME clades, including ANME-1b (40.3%) and ANME-2 (10.0%), during the operation of a methane fed BTF for 248 d. Further research on increasing the bioavailability of methane, process stability, product recovery and large-scale operation is necessary to envision this bioprocess for gas-to-liquid fuel technologies.

References

Ahmaruzzaman, M., 2008. Adsorption of phenolic compounds on low-cost adsorbents: A review. Adv. Colloid Interface Sci. 143, 48–67.

Aljundi, I.H., Belovich, J.M., Talu, O., 2005. Adsorption of lactic acid from fermentation broth and aqueous solutions on Zeolite molecular sieves. Chem. Eng. Sci. 60, 5004–5009.

Alkaya, E., Kaptan, S., Ozkan, L., Uludag-Demirer, S., Demirer, G.N., 2009. Recovery of acids from anaerobic acidification broth by liquid-liquid extraction. Chemosphere 77, 1137–1142.

Anasthas, H.M., Gaikar, V.G., 2001. Adsorption of acetic acid on ion-exchange resins in non-aqueous conditions. React. Funct. Polym. 47, 23–35.

Aylmore, M.G., Muir, D.M., 2001. Thiosulfate leaching of gold - a review. Miner. Eng. 14, 135–174.

Badshah, M., Parawira, W., Mattiasson, B., 2012. Anaerobic treatment of methanol condensate from pulp mill compared with anaerobic treatment of methanol using mesophilic UASB reactors. Bioresour. Technol. 125, 318–327.

Bhattarai, S., Cassarini, C., Rene, E.R., Zhang, Y., Esposito, G., Lens, P.N.L., 2018. Enrichment of sulfate reducing anaerobic methane oxidizing community dominated by ANME-1 from Ginsburg Mud Volcano (Gulf of Cadiz) sediment in a biotrickling filter. Bioresour. Technol. 259, 433–441.

Bousquet, P., Ciais, P., Miller, J.B., Dlugokencky, E.J., Hauglustaine, D.A., Prigent, C., Van der Werf, G.R., Peylin, P., Brunke, E.-G., Carouge, C., Langenfelds, R.L., Lathière, J., Papa, F., Ramonet, M., Schmidt, M., Steele, L.P., Tyler, S.C., White, J., 2006. Contribution of anthropogenic and natural sources to atmospheric methane variability. Nature 443, 439–43.

Brown, D., Shi, J., Li, Y., 2012. Comparison of solid-state to liquid anaerobic digestion of lignocellulosic feedstocks for biogas production. Bioresour. Technol. 124, 379–386.

Cassarini, C., Rene, E.R., Bhattarai, S., Esposito, G., Lens, P.N.L., 2017. Anaerobic oxidation of methane coupled to thiosulfate reduction in a biotrickling filter. Bioresour. Technol. 240, 214–222.

CEPI, 2017. Key Statistics, European pulp and paper industry.

Chen, W.S., Ye, Y., Steinbusch, K.J.J., Strik, D.P.B.T.B., Buisman, C.J.N., 2016. Methanol as an alternative electron donor in chain elongation for butyrate and caproate formation. Biomass and Bioenergy 93, 201–208.

Chowdhury, A., Maranas, C.D., 2015. Designing overall stoichiometric conversions and intervening metabolic reactions. Sci. Rep. 5, 16009.

Coma, M., Vilchez-Vargas, R., Roume, H., Jauregui, R., Pieper, D.H., Rabaey, K., 2016. Product diversity linked to substrate usage in chain elongation by mixed-culture fermentation. Environ. Sci. Technol. 50, 6467–6476.

Conrado, R.J., Gonzalez, R., 2014. Envisioning the bioconversion of methane to liquid fuels. Science. 343, 621–623.

Da Silva, A.H., Miranda, E.A., 2013. Adsorption/desorption of organic acids onto different adsorbents for their recovery from fermentation broths. J. Chem. Eng. Data 58, 1454–1463.

Dessì, P., Jain, R., Singh, S., Seder-Colomina, M., van Hullebusch, E.D., Rene, E.R., Ahammad, S.Z., Carucci, A., Lens, P.N.L., 2016. Effect of temperature on selenium removal from wastewater by UASB reactors. Water Res. 94, 146–154.

Dhamwichukorn, S., Kleinheinz, G.T., Bagley, S.T., 2001. Thermophilic biofiltration of methanol and α-pinene. J. Ind. Microbiol. Biotechnol. 26, 127–133.

Dias, J.C.T., Rezende, R.P., Silva, C.M., Linardi, V.R., 2005. Biological treatment of kraft pulp mill foul condensates at high temperatures using a membrane bioreactor. Process Biochem. 40, 1125–1129.

Djas, M., Henczka, M., 2018. Reactive extraction of carboxylic acids using organic solvents and supercritical fluids: A review. Sep. Purif. Technol. 201, 106–119.

Dorado, A.D., Lafuente, F.J., Gabriel, D., Gamisans, X., 2010. A comparative study based on physical characteristics of suitable packing materials in biofiltration. Environ. Technol. 31, 193–204.

Duan, Z., Mao, S., 2006. A thermodynamic model for calculating methane solubility, density and gas phase composition of methane-bearing aqueous fluids from 273 to 523 K and from 1 to 2000 bar. Geochim. Cosmochim. Acta 70, 3369–3386.

Dufresne, R., Liard, A., Blum, M.S., 2001. Anaerobic treatment of condensates: trial at a kraft pulp and paper mill. Water Environ. Res. 73, 103–109.

Ed Dlugokencky, 2018. ESRL Global Monitoring Division - Global Greenhouse Gas Reference Network, NOAA/ESRL.

Elvidge, C.D., Zhizhin, M., Baugh, K., Hsu, F.C., Ghosh, T., 2016. Methods for global survey of natural gas flaring from visible infrared imaging radiometer suite data. Energies 9.

Eregowda, T., Matanhike, L., Rene, E.R., Lens, P.N.L., 2018. Performance of a biotrickling filter for anaerobic utilization of gas-phase methanol coupled to thiosulphate reduction and resource recovery through volatile fatty acids production. Bioresour. Technol. 263, 591–600.

Espinosa-Ortiz, E.J., Rene, E.R., Guyot, F., van Hullebusch, E.D., Lens, P.N.L., 2017. Biomineralization of tellurium and selenium-tellurium nanoparticles by the white-rot fungus Phanerochaete chrysosporium. Int. Biodeterior. Biodegrad. 124, 258–266.

Fei, Q., Guarnieri, M.T., Tao, L., Laurens, L.M.L., Dowe, N., Pienkos, P.T., 2014. Bioconversion of natural gas to liquid fuel: Opportunities and challenges. Biotechnol. Adv. 32, 596–614.

Fermoso, F.G., Collins, G., Bartacek, J., O'Flaherty, V., Lens, P., 2008. Acidification of methanol-fed anaerobic granular sludge bioreactors by cobalt deprivation: Induction and microbial community dynamics. Biotechnol. Bioeng. 99, 49–58.

Fernández-Nava, Y., Marañón, E., Soons, J., Castrillón, L., 2010. Denitrification of high nitrate concentration wastewater using alternative carbon sources. J. Hazard. Mater. 173, 682–688.

Fernández, M., Ramírez, M., Gómez, J.M., Cantero, D., 2014. Biogas biodesulfurization in an anoxic biotrickling filter packed with open-pore polyurethane foam. J. Hazard. Mater. 264, 529–535.

Foo, K.Y., Hameed, B.H., 2010. Insights into the modeling of adsorption isotherm systems. Chem. Eng. J. 156, 2–10.

Garg, S., Wu, H., Clomburg, J.M., Bennett, G.N., 2018. Bioconversion of methane to C-4 carboxylic acids using carbon flux through acetyl-CoA in engineered *Methylomicrobium buryatense* 5GB1C. Metab. Eng. 48, 175–183.

Ge, X., Yang, L., Sheets, J.P., Yu, Z., Li, Y., 2014. Biological conversion of methane to liquid fuels: status and opportunities. Biotechnol. Adv. 32, 1460–75.

Gribbins, M.J., Loehr, R.C., 1998. Effect of media nitrogen concentration on biofilter performance. J. Air Waste Manag. Assoc. 48, 216–226.

Guerra, D.J.L., Mello, I., Freitas, L.R., Resende, R., Silva, R.A.R., 2014. Equilibrium, thermodynamic, and kinetic of Cr(VI) adsorption using a modified and unmodified bentonite clay. Int. J. Min. Sci. Technol. 24, 525–535.

Hageman, S.P.W., van der Weijden, R.D., Stams, A.J.M., van Cappellen, P., Buisman, C.J.N., 2017. Microbial selenium sulfide reduction for selenium recovery from wastewater. J. Hazard. Mater. 329, 110–119.

Hamdaoui, O., Naffrechoux, E., 2007. Modeling of adsorption isotherms of phenol and chlorophenols onto granular activated carbon. Part I. Two-parameter models and equations allowing determination of thermodynamic parameters. J. Hazard. Mater. 147, 381–394.

Han, W., He, P., Shao, L., Lü, F., 2018. Metabolic interactions of a chain elongation microbiome. Appl. Environ. Microbiol. 84 (22), e01614-e01618.

He, Z., Geng, S., Shen, L., Lou, L., Zheng, P., Xu, X., Hu, B., 2015. The short- and long-term effects of environmental conditions on anaerobic methane oxidation coupled to nitrite reduction. Water Res. 68, 554–562.

Henard, C.A., Smith, H., Dowe, N., Kalyuzhnaya, M.G., Pienkos, P.T., Guarnieri, M.T., 2016. Bioconversion of methane to lactate by an obligate methanotrophic bacterium. Sci. Rep. 6, 1–9.

Hu, Y., Jing, Z., Sudo, Y., Niu, Q., Du, J., Wu, J., Li, Y.Y., 2015. Effect of influent COD/SO_4^{2-} ratios on UASB treatment of a synthetic sulfate-containing wastewater. Chemosphere 130, 24–33.

Hur, D.H., Na, J.G., Lee, E.Y., 2017. Highly efficient bioconversion of methane to methanol using a novel type I Methylomonas sp. DH-1 newly isolated from brewery waste sludge. J. Chem. Technol. Biotechnol. 92, 311–318.

Hwang, I.Y., Lee, S.H., Choi, Y.S., Park, S.J., Na, J.G., Chang, I.S., Kim, C., Kim, H.C., Kim, Y.H., Lee, J.W., Lee, E.Y., 2014. Biocatalytic conversion of methane to methanol as a key step for development of methane-based biorefineries. J. Microbiol. Biotechnol. 24, 1597–1605.

IPCC, 2014. Climate Change 2014: Synthesis Report. Contribution of Working Groups I, II and III to the Fifth Assessment Report of the Intergovernmental Panel on Climate Change, Core Writing Team, R.K. Pachauri and L.A. Meyer.

Iranpour, R., Cox, H.H.J., Deshusses, M.A., Schroeder, E.D., 2005. Literature review of air pollution control biofilters and biotrickling filters for odor and volatile organic compound removal. Environ. Prog. 24, 254–267.

Jones, R.J., Massanet-Nicolau, J., Guwy, A., Premier, G., Dinsdale, R., Reilly, M., 2015. Removal and recovery of inhibitory volatile fatty acids from mixed acid fermentations by conventional electrodialysis. Bioresour. Technol. 189, 279–284.

Kang, T.J., Lee, E.Y., 2016. Metabolic versatility of microbial methane oxidation for biocatalytic methane conversion. J. Ind. Eng. Chem. 35, 8-13.

Khoei, N.S., Lampis, S., Zonaro, E., Yrjälä, K., Bernardi, P., Vallini, G., 2017. Insights into selenite reduction and biogenesis of elemental selenium nanoparticles by two environmental isolates of Burkholderia fungorum. N. Biotechnol. 34, 1–11.

Kim, H.G., Han, G.H., Kim, S.W., 2010. Optimization of lab scale methanol production by methylosinus trichosporium OB3b. Biotechnol. Bioprocess Eng. 15, 476–480.

Kim, S., Deshusses, M.A., 2008. Determination of mass transfer coefficients for packing materials used in biofilters and biotrickling filters for air pollution control-2: Development of mass transfer coefficients correlations. Chem. Eng. Sci. 63, 856–861.

Lai, C.Y., Wen, L.L., Shi, L.D., Zhao, K.K., Wang, Y.Q., Yang, X., Rittmann, B.E., Zhou, C., Tang, Y., Zheng, P., Zhao, H.P., 2016. Selenate and nitrate bioreductions using methane as the electron donor in a membrane biofilm reactor. Envirnmental Sci. Technol. 50, 10179–10186.

Latif, H., Zeidan, A.A., Nielsen, A.T., Zengler, K., 2014. fermenting microorganisms ScienceDirect Trash to treasure : production of biofuels and commodity chemicals via syngas fermenting microorganisms. Curr. Opin. Biotechnol. 27, 79–87.

Lenz, M., Lens, P.N.L., 2009. The essential toxin: The changing perception of selenium in environmental sciences. Sci. Total Environ. 407, 3620–3633.

Levario, T.J., Dai, M., Yuan, W., Vogt, B.D., Nielsen, D.R., 2012. Rapid adsorption of alcohol biofuels by high surface area mesoporous carbons. Microporous Mesoporous Mater. 148, 107–114.

Li, L., Wang, Y., Qi, X., 2015. Adsorption of imidazolium-based ionic liquids with different chemical structures onto various resins from aqueous solutions. RSC Adv. 5, 41352–41358.

Lin, B., 2008. The basics of foul condensate stripping. Krat recovery short course, Florida, USA.

Lin, Y., He, Y., Meng, Z., Yang, S., 2008. Anaerobic treatment of wastewater containing methanol in upflow anaerobic sludge bed (UASB) reactor. Front. Environ. Sci. Eng. China 2, 241–246.

Liu, R., Zhang, P., Zhang, S., Yan, T., Xin, J., Zhang, X., 2016. Ionic liquids and supercritical carbon dioxide: Green and alternative reaction media for chemical processes. Rev. Chem. Eng. 32, 587–609.

López-garzón, C.S., Straathof, A.J.J., 2014. Recovery of carboxylic acids produced by fermentation. Biotechnol. Adv. 32, 873–904.

Mal, J., Nancharaiah, Y. V, van Hullebusch, E.D., Lens, P.N.L., 2017. Biological removal of selenate and ammonium by activated sludge in a sequencing batch reactor. Bioresour. Technol. 229, 11–19.

Martinez, G.A., Rebecchi, S., Decorti, D., Domingos, J.M.B., Natolino, A., Del Rio, D., Bertin, L., Da Porto, C., Fava, F., 2015. Towards multi-purpose biorefinery platforms for the valorisation of red grape pomace: production of polyphenols, volatile fatty acids, polyhydroxyalkanoates and biogas. Green Chem. 18, 261–270.

Meulepas, R.J.W., Jagersma, C.G., Gieteling, J., Buisman, C.J.N., Stams, A.J.M., Lens, P.N.L., 2009. Enrichment of anaerobic methanotrophs in sulfate-reducing membrane bioreactors. Biotechnol. Bioeng. 104, 458–470.

Mudliar, S., Giri, B., Padoley, K., Satpute, D., Dixit, R., Bhatt, P., Pandey, R., Juwarkar, A., Vaidya, A., 2010. Bioreactors for treatment of VOCs and odours - A review. J. Environ. Manage. 91, 1039–1054.

Mueller, T.J., Grisewood, M.J., Nazem-Bokaee, H., Gopalakrishnan, S., Ferry, J.G., Wood, T.K., Maranas, C.D., 2015. Methane oxidation by anaerobic archaea for conversion to liquid fuels. J. Ind. Microbiol. Biotechnol. 42, 391–401.

Nancharaiah, Y.V, Lens, P.N.L., 2015. Ecology and biotechnology of selenium-respiring bacteria. Microbiol. Mol. Biol. Rev. 79, 61–80.

Nancharaiah, Y.V, Lens, P.N.L., 2015. Selenium biomineralization for biotechnological applications. Trends Biotechnol.

Nazaries, L., Murrell, J.C., Millard, P., Baggs, L., Singh, B.K., 2013. Methane, microbes and models: Fundamental understanding of the soil methane cycle for future predictions. Environ. Microbiol. 15, 2395–2417.

Nielsen, D.R., Amarasiriwardena, G.S., Prather, K.L.J., 2010. Predicting the adsorption of second generation biofuels by polymeric resins with applications for in situ product recovery (ISPR). Bioresour. Technol. 101, 2762–2769.

Pan, X.R., Li, W.W., Huang, L., Liu, H.Q., Wang, Y.K., Geng, Y.K., Lam, P.K.-S., Yu, H.Q., 2018. Recovery of high-concentration volatile fatty acids from wastewater using an acidogenesis-electrodialysis integrated system. Bioresour. Technol. 260, 61–67.

Passe-Coutrin, N., Altenor, S., Cossement, D., Jean-Marius, C., Gaspard, S., 2008. Comparison of parameters calculated from the BET and Freundlich isotherms obtained by nitrogen adsorption on activated carbons: A new method for calculating the specific surface area. Microporous Mesoporous Mater. 111, 517–522.

Paulo, P.L., Vallero, M.V.G., Treviño, R.H.M., Lettinga, G., Lens, P.N.L., 2004. Thermophilic (55°C) conversion of methanol in methanogenic-UASB reactors: influence of sulphate on methanol degradation and competition. J. Biotechnol. 111, 79–88.

Pettine, M., McDonald, T.J., Sohn, M., Anquandah, G.A.K., Zboril, R., Sharma, V.K., 2015. A critical review of selenium analysis in natural water samples. Trends Environ. Anal. Chem. 5, 1–7.

Pinelli, D., Molina Bacca, A.E., Kaushik, A., Basu, S., Nocentini, M., Bertin, L., Frascari, D., 2016. Batch and continuous flow adsorption of phenolic compounds from olive mill wastewater: A comparison between nonionic and ion exchange resins. Int. J. Chem. Eng. Article number 9349627.

Pradhan, N., Rene, E.R., Lens, P.N.L., Dipasquale, L., D'Ippolito, G., Fontana, A., Panico, A., Esposito, G., 2017. Adsorption behaviour of lactic acid on granular activated carbon and anionic resins: Thermodynamics, isotherms and kinetic studies. Energies 10, 1–16.

Rattanapan, C., Ounsaneha, W., 2012. Removal of Hydrogen Sulfide Gas using Biofiltration - a Review. Walailak J. 9, 9–18.

Rebecchi, S., Pinelli, D., Bertin, L., Zama, F., Fava, F., Frascari, D., 2016. Volatile fatty acids recovery from the effluent of an acidogenic digestion process fed with grape pomace by adsorption on ion exchange resins. Chem. Eng. J. 306, 629–639.

Rene, E.R., López, M.E., Veiga, M.C., Kennes, C., 2010. Steady- and transient-state operation of a two-stage bioreactor for the treatment of a gaseous mixture of hydrogen sulphide, methanol and α-pinene. J. Chem. Technol. Biotechnol. 85, 336–348.

Reyhanitash, E., Kersten, S.R.A., Schuur, B., 2017. Recovery of volatile fatty acids from fermented wastewater by adsorption. ACS Sustain. Chem. Eng. 5, 9176–9184.

Reyhanitash, E., Zaalberg, B., Kersten, S.R.A., Schuur, B., 2016. Extraction of volatile fatty acids from fermented wastewater. Sep. Purif. Technol. 161, 61–68.

Roghair, M., Hoogstad, T., Strik, D.P.B.T.B., Plugge, C.M., Timmers, P.H.A., Weusthuis, R.A., Bruins, M.E., Buisman, C.J.N., 2018. Controlling ethanol use in chain elongation by CO_2 loading rate. Environ. Sci. Technol. 52, 1496–1506.

Serra, M.C.C., Pessoa, F.L.P., Palavra, A.M.F., 2006. Solubility of methane in water and in a medium for the cultivation of methanotrophs bacteria. J. Chem. Thermodyn. 38, 1629–1633.

Sheets, J.P., Ge, X., Li, Y.F., Yu, Z., Li, Y., 2016. Biological conversion of biogas to methanol using methanotrophs isolated from solid-state anaerobic digestate. Bioresour. Technol. 201, 50–57.

Sheha, R.R., El-Shazly, E.A., 2010. Kinetics and equilibrium modeling of Se(IV) removal from aqueous solutions using metal oxides. Chem. Eng. J. 160, 63–71.

Soda, S., Kashiwa, M., Kagami, T., Kuroda, M., Yamashita, M., Ike, M., 2011. Laboratory-scale bioreactors for soluble selenium removal from selenium refinery wastewater using anaerobic sludge. Desalination 279, 433–438.

Soo, V.W.C., McAnulty, M.J., Tripathi, A., Zhu, F., Zhang, L., Hatzakis, E., Smith, P.B., Agrawal, S., Nazem-Bokaee, H., Gopalakrishnan, S., Salis, H.M., Ferry, J.G., Maranas, C.D., Patterson, A.D., Wood, T.K., 2016. Reversing methanogenesis to capture methane for liquid biofuel precursors. Microb. Cell Fact. 15, 11.

Staicu, L.C., Morin-Crini, N., Crini, G., 2017. Desulfurization: Critical step towards enhanced selenium removal from industrial effluents. Chemosphere 172, 111–119.

Stolaroff, J.K., Bhattacharyya, S., Smith, C.A., Bourcier, W.L., Cameron-Smith, P.J., Aines, R.D., 2012. Review of methane mitigation technologies with application to rapid release of methane from the Arctic. Environ. Sci. Technol. 46, 6455–69.

Taher, E., Chandran, K., 2013. High-rate, high-yield production of methanol by ammonia-oxidizing bacteria. Environ. Sci. Technol. 47, 3167–3173.

Takada, T., Hirata, M., Kokubu, S., Toorisaka, E., Ozaki, M., Hano, T., 2008. Kinetic study on biological reduction of selenium compounds. Process Biochem. 43, 1304–1307.

Tan, L.C., Calix, E.M., Rene, E.R., Nancharaiah Y. V, van Hullebusch, E.D., Lens, P.N.L., 2018. Amberlite IRA-900 ion exchange resin for the sorption of selenate and sulfate: equilibrium, kinetic, and regeneration studies. J. Environ. Eng. 144.11: 04018110.

Tan, L.C., Nancharaiah, Y. V, van Hullebusch, E.D., Lens, P.N.L., 2016. Selenium: environmental significance, pollution, and biological treatment technologies. Biotechnol. Adv. *34*, 886-907.

Tran, H., Vakkilainnen, E.K., 2012. The Kraft Chemical Recovery Process. TAPPI Kraft Recover. Course 1–8.

TRI Program, 2017. 2017 TRI Factsheet for Chemical METHANOL.

Tugtas, A.E., 2011. Fermentative organic acid production and separation. Fen Bilim. Derg. 23, 70–78.

Uslu, H., Bayazit, S., 2010. Adsorption equilibrium data for acetic acid and glycolic acid onto Amberlite 298, 1295–1299.

Wang, C., Li, Q., Wang, D., Xing, J., 2014. Improving the lactic acid production of *Actinobacillus succinogenes* by using a novel fermentation and separation integration system. Process Biochem. 49, 1245–1250.

Wang, D., Wang, Y., Liu, Y., Ngo, H.H., Lian, Y., Zhao, J., Chen, F., Yang, Q., Zeng, G., Li, X., 2017. Is denitrifying anaerobic methane oxidation-centered technologies a solution for the sustainable operation of wastewater treatment Plants? Bioresour. Technol. 234, 456–465.

Wang, Y., Wang, D., Yang, Q., Zeng, G., Li, X., 2017. Wastewater opportunities for denitrifying anaerobic methane oxidation. Trends Biotechnol. 35, 799–802.

Wani, A.H., Lau, A.K., Branion, R.M.R., 1999. Biofiltration control of pulping odors - hydrogen sulfide: Performance, macrokinetics and coexistence effects of organo-sulfur species. J. Chem. Technol. Biotechnol. 74, 9–16.

Weijma, J., Chi, T.M., Hulshoff Pol, L.W., Stams, A.J.M., Lettinga, G., 2003. The effect of sulphate on methanol conversion in mesophilic upflow anaerobic sludge bed reactors. Process Biochem. 38, 1259–1266.

Weijma, J., Stams, A.J.M., 2001. Methanol conversion in high-rate anaerobic reactors. Water Sci. Technol. 44, 7–14.

Yang, L., Ge, X., Wan, C., Yu, F., Li, Y., 2014. Progress and perspectives in converting biogas to transportation fuels. Renew. Sustain. Energy Rev. 40, 1133–1152.

Yang, S.T., White, S.A., Hsu, S.T., 1991. Extraction of carboxylic acids with tertiary and quaternary amines: effect of pH. Ind. Eng. Chem. Res. 30, 1335–1342.

Yousuf, A., Bonk, F., Bastidas-Oyanedel, J.R., Schmidt, J.E., 2016. Recovery of carboxylic acids produced during dark fermentation of food waste by adsorption on Amberlite IRA-67 and activated carbon. Bioresour. Technol. 217, 137–140.

Zhang, S., Wang, X., Li, J., Wen, T., Xu, J., Wang, X., 2014. Efficient removal of a typical dye and Cr(vi) reduction using N-doped magnetic porous carbon. RSC Adv. 4, 63110–63117.

Zhang, Y., Angelidaki, I., 2015. Bioelectrochemical recovery of waste-derived volatile fatty acids and production of hydrogen and alkali. Water Res. 81, 188-195

Zhang, Y., Henriet, J.P., Bursens, J., Boon, N., 2010. Stimulation of in vitro anaerobic oxidation of methane rate in a continuous high-pressure bioreactor. Bioresour. Technol. 101, 3132–3138.

CHAPTER 3

Selenate and thiosulfate reduction using methanol as electron donor

This chapter will be modified and published as:

Eregowda, T., Rene, E.R., Matanhike, L., Lens, P.N.L., 2019. Selenate and thiosulfate reduction using methanol as electron donor. Manuscript submitted to Environmental Science and Pollution Research.

Abstract

Anaerobic bioconversion of methanol was tested in the presence of selenate (SeO_4^{2-}), thiosulfate ($S_2O_3^{2-}$) and sulfate (SO_4^{2-}) as electron acceptors. Complete SeO_4^{2-} reduction occurred at COD:SeO_4^{2-} ratios of 12 and 30, whereas ~ 83% reduction occurred when the COD:SeO_4^{2-} ratio was 6. Methane production did not occur at the three COD:SeO_4^{2-} ratios investigated. Up to 10.1 and 30.9% of $S_2O_3^{2-}$ disproportionated to SO_4^{2-} at COD:$S_2O_3^{2-}$ ratios of 1.2 and 2.25, respectively, and > 99% reduction was observed at both ratios. The presence of $S_2O_3^{2-}$ lowered the methane production by 73.1% at a COD:$S_2O_3^{2-}$ ratio of 1.2 compared to the control. This study showed that biogas production is not preferable for SeO_4^{2-} and $S_2O_3^{2-}$ rich effluents such as pulp and paper, and petroleum refining wastewaters, whereas volatile fatty acids (VFA) production could be a potential resource recovery option for these wastewaters.

Keywords: Selenate reduction, thiosulfate reduction, thiosulfate disproportionation, VFA, methane, methylotrophy

3.1. Introduction

Methanol is a common volatile organic compound (VOC) present in the effluents of process industries such as the pulp and paper, paint manufacturing and petroleum refineries. In the year 2017, up to 55,377 tonnes of methanol were emitted to the atmosphere in the United States alone and at least 65% of the total emission was from the Kraft mills of the pulp and paper industries (TRI Program, 2017). Several studies have demonstrated the utilization of methanol from methanol rich liquid effluents such as the utilization of methanol rich gaseous emissions for volatile fatty acid (VFA) production (Eregowda et al., 2018) and Kraft condensates from the Kraft pulping industries for methane production (Badshah et al., 2012). Anaerobic utilization of methanol rich effluents provides an opportunity for resource recovery through the production of methane or waste-derived VFA (Eregowda et al., 2018).

Methylotrophic microorganisms utilize methanol as the substrate for methane and VFA production. They play a key role in the anaerobic treatment of effluents rich in alcohols and sugars (Large, 1983). The effect of several inhibitory factors on the anaerobic digestion of methanol, including high concentrations of ammonia, sulfide and heavy metals have been reported in the literature (Chen et al., 2008). However, to the best of our knowledge, the effects of oxyanions such as selenate (SeO_4^{2-}) and thiosulfate ($S_2O_3^{2-}$) have not yet been reported. Selenium (Se) is a chalcogen (group 16 of the periodic table) and a trace element naturally present in bed rocks. Although Se is an essential element (at <40 µg/d), an intake >400 µg/d can be toxic to the living organisms (Bleiman and Mishael, 2010). Se shares similar physico-chemical properties with sulfur and is closely associated with sulfur-containing minerals, pyrites and fossil fuel sources (Mehdi et al., 2013). Due to its chalcophilic nature, i.e. strong affinity with sulfur (Tan et al., 2016), Se compounds have a close analogy with sulfur compounds (Mehdi et al., 2013) and the biochemical similarities of Se and sulfur allow their exchange by sulfate and selenate reducing organisms (Hockin and Gadd, 2003a).

Wastewaters from industries such as oil refining or lead, gold and Se mining can, respectively, contain up to 5, 7, 33 and 620 mg/L of Se oxyanions (SeO_x^{2-}) (Pettine et al., 2015). Countries like South Korea, Canada and Japan have stringent effluent discharge limits for SeO_x^{2-}, in the range of 0.001-0.1 mg/L (Tan et al., 2016). Moreover, the oxyanions of sulfur such as sulfate (SO_4^{2-}) and thiosulfate ($S_2O_3^{2-}$) are common constituents of chemical pulping, petroleum refining and mining industry effluents (Speece, 1983). For example, petroleum industries generate $S_2O_3^{2-}$ (0.001-50 g/L) due to the partial oxidation of sulfide in gas scrubbers

(González-Sánchez et al., 2008). Additionally, $S_2O_3^{2-}$ is one of the main intermediates formed during biological SO_4^{2-}/sulfite (SO_3^{2-}) reduction, with its reduction to sulfide being the rate-limiting step (Qian et al., 2017). Although the bioreduction of SO_4^{2-} has been extensively studied and applied in full-scale systems [18-20], $S_2O_3^{2-}$ reduction has only been reported in a few studies (Cassarini et al., 2017; González-Sánchez et al., 2008; Jørgensen et al., 1991; Qian et al., 2017). The objective of this study was, therefore, to examine the effect of SeO_4^{2-}, $S_2O_3^{2-}$ or SO_4^{2-} on methanol utilization for VFA or methane production.

3.2. Materials and methods

3.2.1. Source of biomass and medium composition

The activated sludge (volatile suspended solids content: 2.86 ± 0.02 g/L) used as the inoculum was collected from a full-scale domestic wastewater treatment plant located in Harnaschpolder (the Netherlands). The characteristics of the sludge in terms of alkali, alkaline earth and heavy metals were as described in Matanhike (2017). The sludge consisted of 69 mg/L sulfate and Se was below detection limit (< 0.2 µg/L). In terms of trace elements, the sludge comprised of (in µg/L) Sodium (557), Magnesium (263), Aluminium (179), Potassium (352), Calcium (147), Manganese (6,5), Iron (296) Copper (7.7) and Nickel (1.1). DSM 63 media modified to eliminate or replace sulfate salts with chlorides was used as the synthetic wastewater and consisted of 0.5 g K_2HPO_4, 1.0 g NH_4Cl, 0.1 g $CaCl_2{\cdot}2H_2O$, 1.0 g $MgCl_2{\cdot}6H_2O$, 1 g yeast extract and 0.5 mg resazurin as the redox indicator in 980 mL of MilliQ water, autoclaved at 121 °C for 20 min . 10 mL each of sodium thioglycolate solution (10 g/L) and ascorbic acid solution (10 g/L) were added and the pH was maintained at 7.5 (± 0.1) (DSM 63; Eregowda et al., 2018).

3.2.2. Experimental design

The effects of SeO_4^{2-} and $S_2O_3^{2-}$ on methylotrophic bacteria were tested in batch bottles at different concentrations of methanol (electron donor) and the individual electron acceptors (SeO_4^{2-}, $S_2O_3^{2-}$ or SO_4^{2-}) as described in ***Table 3.1***. The incubations were carried out in 300 mL airtight serum bottles, in duplicate, with a working volume of 200 mL. 15 mL of N_2 purged settled activated sludge (volatile suspended solids: 8.5 ± 0.5 g/L) was added as the inoculum to the bottles (pH of 7.5). Negative and positive controls included batches with methanol only, SeO_4^{2-}/$S_2O_3^{2-}$/SO_4^{2-} only and no biomass.

The bottles containing the inoculum, methanol and the electron acceptor in synthetic wastewater media were purged with N_2 and incubated on an orbital shaker (Cole-Parmer, Germany) at 170 rpm and 22 (± 2) °C. The initial methanol concentration was 200 mg/L for all the batches with SeO_4^{2-}. In order to maintain the COD:$S_2O_3^{2-}$ and COD:SO_4^{2-} ratios of 1.2 and 2.25, the methanol concentration was 600 mg/L for the batches with 400 mg/L SO_4^{2-} or $S_2O_3^{2-}$ and 800 mg/L for the batches with 1000 mg/L SO_4^{2-} or $S_2O_3^{2-}$. The batch incubations were terminated after the complete utilization of methanol, which occurred in 14, 19, 38 and 45 d, respectively, in the batches with 0, 10, 25 and 50 mg/L SeO_4^{2-} and in 35 and 22 d in the batches with the COD:SO_4^{2-} or COD:$S_2O_3^{2-}$ ratios of 1.2 and 2.25, respectively.

Liquid samples to measure the pH, SeO_4^{2-}, $S_2O_3^{2-}$, SO_4^{2-}, methanol and VFA (acetate, propionate, isobutyrate, butyrate, isovalerate, and valerate) concentrations were filtered using 0.45 μm cellulose acetate syringe filters (Sigma Aldrich, USA) and stored at 4 °C before analysis. The sampling was carried out in duplicate and the data presented in the plots are the average values (average of the duplicate sampling of the duplicate batch systems). The reduction rates (mg/L.d) of SeO_4^{2-}, $S_2O_3^{2-}$, and SO_4^{2-} were calculated as a fraction of the difference between their respective initial and final concentrations to the duration of the batch incubations. The methanol utilization rate (mg/L.d) was calculated as the fraction of concentration to the duration of batch incubations, whereas the production of VFA (mg/L) was cumulatively calculated as the molar acetate equivalent (mg/L).

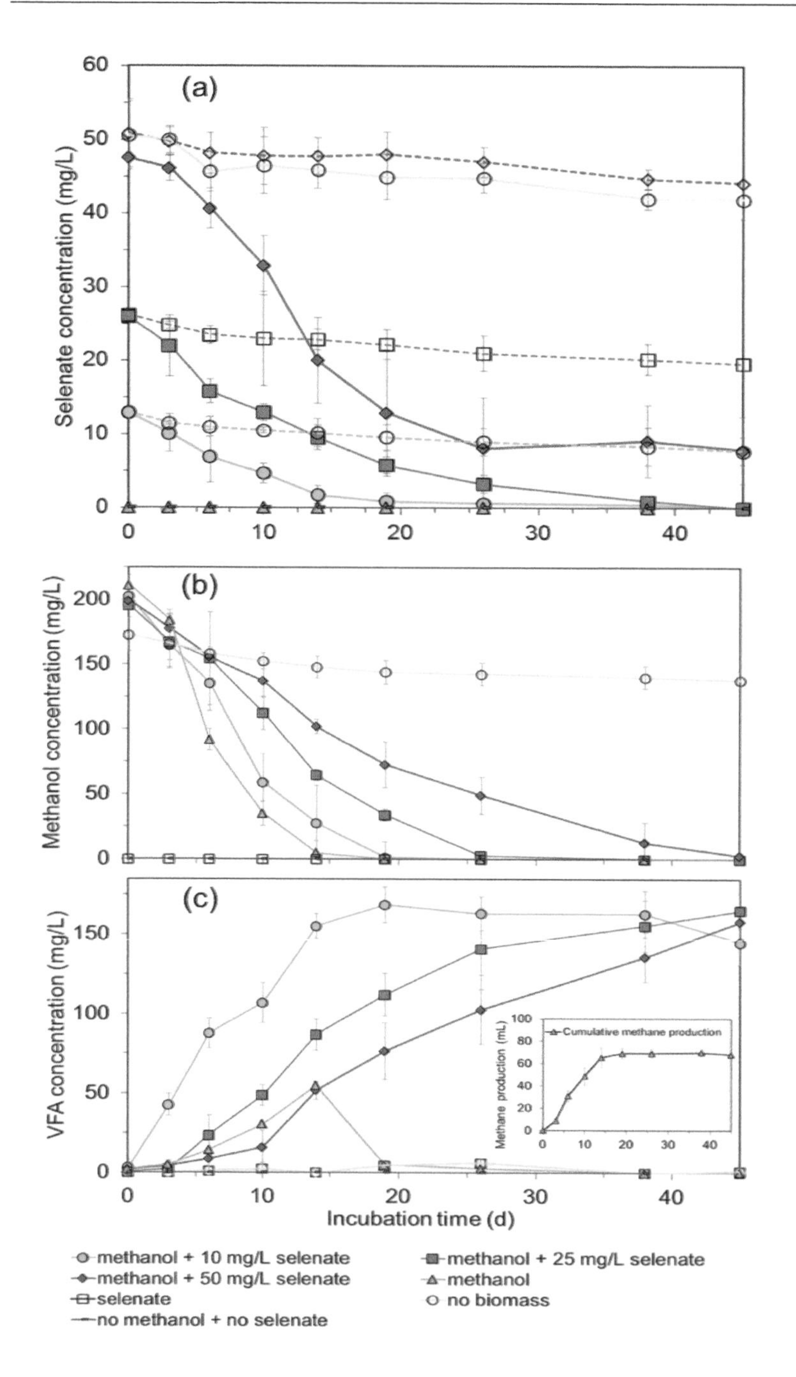

Figure 3.1: *(a) Selenate concentration profiles in the batches with 10 (green), 25 (blue) and 50 (red) mg/L selenate. Dotted lines indicate the selenate profile in the batches without the carbon source (methanol), (b) Methanol concentration profiles and (c) VFA (acetate + propionate + isobutyrate + butyrate + isovalerate + valerate) concentration profiles as*

molar acetate equivalent in batches with 0 (yellow), 10 (green), 25 (blue) and 50 (red) mg/L SeO_4^{2-} and methane production profile (embedded profile) in black, in batches with only methanol.

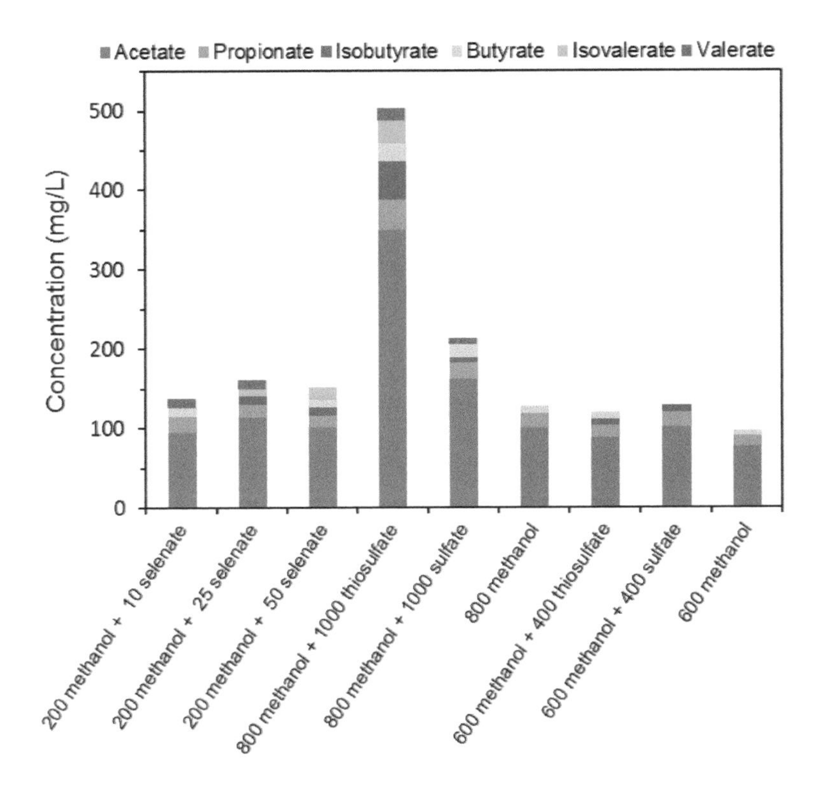

***Figure 3.2:** Composition of individual VFA, i.e. acetate, propionate, isobutyrate, butyrate, isovalerate and valerate under different conditions tested.*

3.2.3. Analytical methods

The pH was analysed using a pH probe (Sen-Tix WTW, the Netherlands) and a pH meter (pH/mV transmitter, DO9785T Delta OHM, the Netherlands). SeO_4^{2-}, $S_2O_3^{2-}$, SO_4^{2-} concentrations were analysed using ion chromatography (ICS-1000, Dionex) as described by Tan et al. (Tan et al., 2018b). Acetate, propionate, isobutyrate, butyrate, isovalerate and valerate concentrations were analysed using gas chromatography (GC) (Varian 430-GC, Varian Inc., USA) as described by Eregowda et al. (Eregowda et al., 2018) and cumulatively expressed as molar acetate equivalent (mg/L). The methanol concentration was measured using a Bruker Scion 456 GC (SCION Instruments, the Netherlands) as described by

Eregowda et al. (Eregowda et al., 2018), while methane was measured using a VARIAN 3800 GC (the Netherlands) as described by Bhattarai et al. (Bhattarai et al., 2018a).

3.3. Results

3.3.1. Selenate reduction and its effect on methanol utilization

Complete SeO_4^{2-} reduction was achieved within 19 and 38 d at the rate of 0.69 and 0.685 mg/L.d, respectively, in the batches with 10 and 25 mg/L SeO_4^{2-} (***Figure 3.1a***). In the batches with 50 mg/L SeO_4^{2-}, ~83% reduction occurred at a rate of 1.52 mg/L.d in 26 d and further reduction of SeO_4^{2-} did not occur. In the incubations without methanol, 8-11% SeO_4^{2-} reduction occurred, while in the incubations without biomass, only ~14% removal of SeO_4^{2-} occurred during the initial 6 d of incubation.

The batch incubations were terminated after the complete utilization of methanol in the batches with 0, 10, 25 and 50 mg/L SeO_4^{2-}, respectively, at the rate of 15.1, 10.6, 7.5 and 4.4 mg/L.d. Along with acetate, 10-35 mg/L of propionate, isobutyrate, butyrate, isovalerate and valerate were produced in the different batch incubations (***Figure 3.1c***). The composition of individual VFA for the total VFA as molar acetate equivalent (mg/L) is shown in ***Figure 3.2***. Methane production in the batches with SeO_4^{2-} was negligible (< 3% mL). In the batches without SeO_4^{2-}, a cumulative methane production of ~68 mL occurred.

3.3.2. $S_2O_3^{2-}$ and SO_4^{2-} reduction using methanol as electron donor

At COD:$S_2O_3^{2-}$ ratios of 1.2 and 2.25, the $S_2O_3^{2-}$ reduction was 87.4 and 84.1% at a reduction rate of 23.4 and 24 mg/L.d, respectively (***Figure 3.3***). In the batches with a COD:$S_2O_3^{2-}$ ratio of 2.25, the $S_2O_3^{2-}$ concentration profiles showed an increasing trend after 14 d of incubation. Nearly 10 and 30 % of $S_2O_3^{2-}$ disproportionated to SO_4^{2-} in the batches with COD:$S_2O_3^{2-}$ ratios of 1.2 and 2.25, respectively. Complete $S_2O_3^{2-}$ reduction did not occur in the batches with a COD:$S_2O_3^{2-}$ ratio of 2.25.

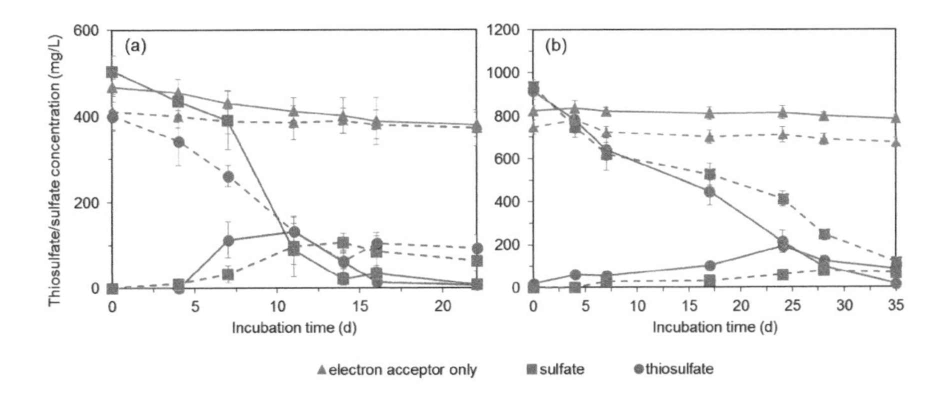

Figure 3.3: *Thiosulfate and sulfate concentration profiles in the batches with a COD:*SO_4^{2-} *(or* $S_2O_3^{-}$*) ratio of (a) 2.25 and (b) 1.2. Dotted lines indicate the batch incubations with thiosulfate as the electron acceptor, while full lines represent the batch incubations with sulfate.*

In the batches with SO_4^{2-} as the electron acceptor, at COD:SO_4^{2-} ratios of 1.2 and 2.25, a reduction of 98.3 and 96.7% SO_4^{2-} occurred at a rate of 25.8 and 34.6 mg/L/d, respectively. Meanwhile, $S_2O_3^{2-}$ was formed as an intermediate (***Figure 3.3***) and ~18.2 and 22.3% SO_4^{2-} was converted to $S_2O_3^{2-}$, that was eventually reduced by the biomass. Overall, the intermediate formation of $S_2O_3^{2-}$ in the batches with SO_4^{2-} or SO_4^{2-} in the batches with $S_2O_3^{2-}$ was higher at a COD:SO_4^{2-} (or $S_2O_3^{2-}$) ratio of 2.25 compared to a COD:SO_4^{2-} (or $S_2O_3^{2-}$) ratio of 1.2.

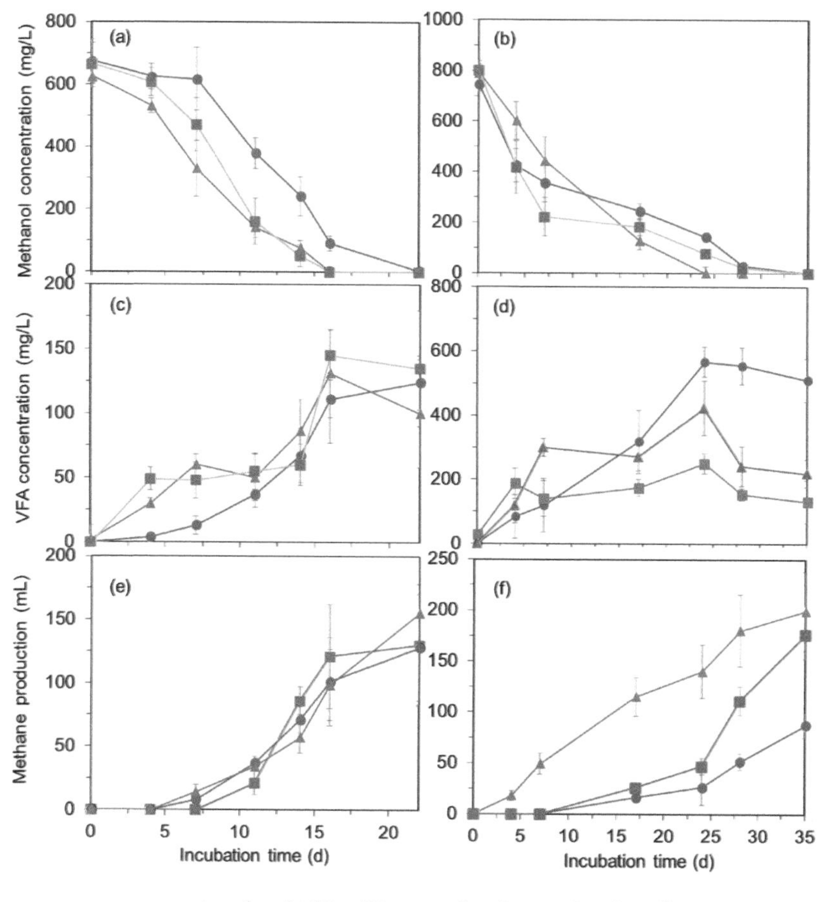

Figure 3.4: *Time course profiles of (a, c, e) methanol, VFA and methane concentration, respectively, in batches with COD:*SO_4^{2-} *or COD:*$S_2O_3^{2-}$ *ratio of 1.2, (b, d, f) methanol, VFA and methane concentration, respectively, in batches with COD:*SO_4^{2-} *or COD:*$S_2O_3^{2-}$ *ratio of 2.25. The red, green and blue lines indicate the batches with thiosulfate, sulfate and no electron acceptor, respectively.*

Table 3.1. *Selenate, sulfate and thiosulfate reduction, methanol utilization and VFA production rates in batch tests using different COD:*SeO_4^{2-} *(or* $S_2O_3^{2-}$ *or* SO_4^{2-}*) ratios.*

COD:SeO_4^{2-} (or $S_2O_3^{2-}$ or SO_4^{2-}) ratio	**Na**	**COD:SeO_4^{2-} = 30**	**COD:SeO_4^{2-} = 12**	**COD:SeO_4^{2-} = 6**	**COD: $S_2O_3^{2-}$ or SO_4^{2-} = 1.2**			**COD:$S_2O_3^{2-}$ or SO_4^{2-} = 2.25**		
Electron acceptor	Control	Selenate			Thiosulfate	Sulfate	Control	Thiosulfate	Sulfate	Control
Electron acceptor concentration (mg/L)	0	10	25	50	1000	1000	0	400	400	0
Methanol concentration (mg/L)	200	200	200	200	800	800	800	600	600	600
Duration (d)*	14	19	38	45⁺	35	35	21	22	22	16
Methanol removal rate (mg/L.d)	15.1	10.6	7.5	4.4	26.6	32.1	39.2	30.6	41.6	44.8
Electron acceptor removal rate (mg/L.d)	Na	0.69	0.68	1.52⁺	23.4	25.8	na	24	34.6	na
VFA production (mg/L)#	-	162.5	155.3	157.9	874	804	300	243	312	376
Cumulative methane production (mL)	68	in	in	in	87	176	324	201	237	287

Note:

[na] Not applicable

[in] Insignificant (<5% *v/v*)

* Number of days until methanol was completely consumed

⁺ Selenate concentration did not reduce below 8.12 mg/L

Calculated as molar acetate equivalent

The methanol utilization rate in the batches with $S_2O_3^{2-}$ was lower compared to the batches with SO_4^{2-} at the tested COD:SO_4^{2-} or COD:$S_2O_3^{2-}$ ratios (***Table 3.1***). The carbon mass balance analysis showed that up to 92% of methanol was recovered as VFA and methane. Up to 510 and 124 mg/L VFA was produced in the batches at a COD:$S_2O_3^{2-}$ ratio of, respectively, 1.2 and 2.25.

The methane production showed a linearly increasing trend at the COD:SO_4^{2-} or COD:$S_2O_3^{2-}$ ratios of 1.2 and 2.25, indicating that the VFA produced from methanol were eventually converted to methane (***Figure 3.4e and 3.4f***). In the batches with $S_2O_3^{2-}$, at a COD:$S_2O_3^{2-}$ ratio of 1.2, the methane production was low (87 mL) compared to the batches with SO_4^{2-} (176 mL) or without electron acceptor (196 mL). Furthermore, comparing the methanol utilization for VFA production at different COD:SO_4^{2-} or COD:$S_2O_3^{2-}$ ratios, it is evident that the acetogenic activity was higher at a ratio of 1.2 (***Figure 3.4c and 3.4e***) while the methanogenic activity was higher at a ratio of 2.25 (***Figure 3.4d and 3.4f***), clearly indicating that the methanogens were negatively affected at a lower COD:$S_2O_3^{2-}$ ratio (1.2).

3.4. Discussion

3.4.1. Methanogenesis versus acetogenesis of methanol

Anaerobic mineralization of methanol mainly involves methanogens that directly convert methanol to methane, methylotrophic acetogens that convert methanol to acetate or butyrate and acetoclastic methanogens that further convert acetate and butyrate to methane [26-29]. This study showed that methylotrophy through methanogenesis was completely inhibited by 10 mg/L SeO_4^{2-} (***Figure 3.1c***). At a low COD:$S_2O_3^{2-}$ ratio of 1.2, methanogenesis was 73.1% compared to the control during the study period. The methanol was utilized by acetogens and up to 93% was converted to VFA.

The inhibition of methanogens is commonly induced by SO_4^{2-} (or $S_2O_3^{2-}$) reduction, especially at a lower COD:SO_4^{2-} ratio because: (i) the sulfate reducing bacteria (SRB) compete for the substrate, and (ii) the population of methanogenic bacteria declines due to inhibition by sulfide (Chen et al., 2014). The inhibition of methanogens was stronger during $S_2O_3^{2-}$ reduction, resulting in a ~49.3% decrease of the methane production when compared to the amount of methane produced in the presence of SO_4^{2-} (***Figure 3.3e, 3.3f***). The outcome of the competition between SRB and methanogens is attributed to the COD:SO_4^{2-} (or $S_2O_3^{2-}$) ratio, with acetoclastic methanogens dominating at a ratio > 2.7 and SRB dominating when the ratio is $<$ 1.7 (Chen et al., 2008).

SeO_x^{2-} and $S_2O_3^{2-}$ rich industrial effluents are not ideal candidates for biogas production (***Figure 3.1c, 3.3e and 3.3f***). However, the production of waste-derived VFA is a potential resource recovery option to deal with wastes containing COD, SeO_4^{2-}, $S_2O_3^{2-}$ and/or SO_4^{2-}. The VFA could be recovered using physico-chemical processes such as adsorption, solvent extraction, membrane separation, distillation and electrodialysis (Lee et al., 2014). Future studies should focus on aspects such as VFA production in semi-industrial scale bioreactors using real industrial S/Se-containing wastewater as well as separation and recovery of the desired VFA from the bioreactor, and metagenomic sequencing studies can help in determining the microbial community structure and deducting the VFA producing pathways in anaerobic methylotrophy.

3.4.2. Thiosulfate and sulfate reduction

In methanol and SO_4^{2-} fed upflow anaerobic sludge blanket (UASB) reactors, SRB can utilize methanol as electron donor for the reduction of SO_4^{2-} to H_2S (Liamleam and Annachhatre, 2007) and exist syntrophically with acetogens, since SRB provide 2 moles of bicarbonate per mole of $S_2O_3^{2-}$ or 1.25 moles of bicarbonate per mole of SO_4^{2-} reduced. SRB have a thermodynamic advantage over both acetogens and methanogens and a higher affinity for methanol as the substrate than methanogens (Eregowda et al., 2018).

Desulfomicrobium and *Desulfobulbus* are among the dominant SRB groups in the sulfite and SO_4^{2-} reducing UASB reactors (Qian et al., 2015), as well as $S_2O_3^{2-}$ reducing biomass. Bacterial species from these two genera can use lactate, acetate, pyruvate, glycerol and methanol as the electron donor and carbon source (Eregowda et al., 2018; Qian et al., 2017) to facilitate direct $S_2O_3^{2-}$ reduction to sulfide or disproportionation to SO_4^{2-} and sulfide, followed by reduction of SO_4^{2-} to sulfide (Cassarini et al., 2017). *Desulfobacter postgatei, Desulfobulbus propionicus* and *Desulfovibrio intestinalis* were indeed the most commonly detected SRB species in the activated sludge collected from the Harnaschpolder wastewater treatment plant (Roest et al., 2015), which was also the source of inoculum for this study.

Cypionka et al. [32] studied the $S_2O_3^{2-}$ disproportionation pathway and showed that the $S_2O_3^{2-}$ reduction to SO_4^{2-} occurs through the formation of sulfite and elemental sulfur (S^0) as the intermediates (Cypionka et al., 1998). The slower reduction of $S_2O_3^{2-}$ compared to SO_4^{2-} in this study could be due to the fact that reduction of $S_2O_3^{2-}$ to sulfide is the rate-limiting step during biological SO_4^{2-} or sulfite reduction (Qian et al., 2017).

3.4.3. Selenate reduction

Previous studies on elucidating the biochemical pathways for the anoxic reduction of SeO_4^{2-} showed that SeO_4^{2-} reduction can occur through dissimilatory or detoxification mechanisms [33,34]. Dissimilatory bioreduction of both SeO_4^{2-} and selenite (SeO_3^{2-}) is mediated by a periplasmic reductase system and the majority of the SeO_4^{2-}/SeO_3^{2-} reduction to Se^0 occurs either in the periplasm or outside the cell envelope. Furthermore, SeO_3^{2-} reduction is also catalysed by other terminal reductases, such as nitrite reductase or sulfite reductase (Y V Nancharaiah and Lens, 2015; Tan et al., 2016).

At higher SeO_4^{2-} concentrations, a longer incubation time was required for the complete utilization of methanol (***Figure 3.3***). It is noteworthy to mention that the methanogens were inhibited to a greater extent compared to the acetogens. The mechanism of inhibition in the presence of SeO_4^{2-} can be due to the generation of free radical species that induce DNA damage or Se reactivity with thiols that affect the function of the DNA repair proteins (Letavayov, 2006). Furthermore, the substitution of seleno-amino acid in the place of its sulfur analogue could also influence the intercellular enzyme activity (Lenz and Lens, 2009).

SeO_x^{2-} reduction has been used in environmental technology for Se removal, i.e. aerobic SeO_3^{2-} (12.7 mg/L) reduction in continuous flow activated sludge reactor (Jain et al., 2016), anaerobic SeO_4^{2-} (7.1 mg/L) co-reduction with nitrate in an upflow anaerobic sludge blanket reactor (Dessì et al., 2016), and anaerobic SeO_4^{2-} (9.15 mg/L) reduction by denitrifying anaerobic methane oxidizing biofilm in a membrane biofilm reactor (Luo et al., 2018). This is, however, the first study to examine the use of methanol as suitable electron donor for the treatment of high concentration (up to 50 mg/L) SeO_x^{2-} synthetic effluent. Studies on the reduction of selenate at high concentration is relevant for the effluents from the selenium, gold and lead mining industries that contain up 620, 33 and 7 mg/L of SeO_x^{2-}, respectively (Pettine et al., 2015).

3.5. Conclusions

Batch studies on anaerobic methanol utilization in the presence of SeO_4^{2-} and $S_2O_3^{2-}$ showed that the presence of SeO_4^{2-} (10 mg/L) completely inhibited methanogens and the methanol was utilized by acetogens for VFA production. In the batch incubations with $S_2O_3^{2-}$, the disproportionation of $S_2O_3^{2-}$ to SO_4^{2-} increased with the COD:$S_2O_3^{2-}$ ratio. The presence of $S_2O_3^{2-}$ at a COD:$S_2O_3^{2-}$ ratio of 1.2 induced a 73.1% reduction of the methane production.

Acknowledgements

The authors thank the lab staff of UNESCO-IHE for their analytical support and the staff from Harnaschpolder WWTP (Delft, the Netherlands) for providing the activated sludge. This work was supported by the Marie Skłodowska-Curie European Joint doctorate (EJD) in Advanced Biological Waste-to-Energy Technologies (ABWET) funded from Horizon 2020 under the grant agreement no. 643071.

References

Badshah M, Parawira W, Mattiasson B. Anaerobic treatment of methanol condensate from pulp mill compared with anaerobic treatment of methanol using mesophilic UASB reactors. Bioresour Technol 2012; 125:318-27.

Bhattarai S, Cassarini C, Naangmenyele Z, Rene ER, Gonzalez-Gil G, Esposito G, et al. Microbial sulfate-reducing activities in anoxic sediment from Marine Lake Grevelingen: Screening of electron donors and acceptors. Limnology 2018; 19:31-41.

Bleiman N, Mishael YG. Selenium removal from drinking water by adsorption to chitosan-clay composites and oxides: Batch and columns tests. J Hazard Mater 2010; 183:590-595.

Cao J, Zhang G, Mao ZS, Li Y, Fang Z, Yang C. Influence of electron donors on the growth and activity of sulfate-reducing bacteria. Int J Miner Process 2012;106-109:58-64.

Cassarini C, Rene ER, Bhattarai S, Esposito G, Lens PNL. Anaerobic oxidation of methane coupled to thiosulfate reduction in a biotrickling filter. Bioresour Technol 2017; 240:214-222.

Cerrillo M, Morey L, Viñas M, Bonmatí A. Assessment of active methanogenic archaea in a methanol-fed upflow anaerobic sludge blanket reactor. Appl Microbiol Biotechnol 2016; 100:10137-10146.

Chen JL, Ortiz R, Steele TWJ, Stuckey DC. Toxicants inhibiting anaerobic digestion: A review. Biotechnol Adv 2014; 32:1523-1534.

Chen Y, Cheng JJ, Creamer KS. Inhibition of anaerobic digestion process : A review. Bioresour

Technol 2008; 99:4044-4064.

Cypionka H, Smock AM, Böttcher ME. A combined pathway of sulfur compound disproportionation in *Desulfovibrio desulfuricans*. FEMS Microbiol Lett 1998; 166:181-186.

Dessì P, Jain R, Singh S, Seder-Colomina M, van Hullebusch ED, Rene ER, et al. Effect of temperature on selenium removal from wastewater by UASB reactors. Water Res 2016; 94:146–154.

DSM 63, Leibniz Institut DSMZ-Deutsche Sammlung von Mikroorganismen und Zellkulturen GmbH ; Curators of the DSMZ.

Eregowda T, Matanhike L, Rene ER, Lens PNL. Performance of a biotrickling filter for anaerobic utilization of gas-phase methanol coupled to thiosulphate reduction and resource recovery through volatile fatty acids production. Bioresour Technol 2018; 263:591-600.

González-Sánchez A, Meulepas R, Revah S. Sulfur formation and recovery in a thiosulfate-oxidizing bioreactor. Environ Technol 2008; 29:847-853.

Hockin SL, Gadd GM. Linked redox precipitation of sulfur and selenium under anaerobic conditions by sulfate-reducing bacterial biofilms 2003; 69:7063–7072.

Jain R, Matassa S, Singh S, van Hullebusch ED, Esposito G, Lens PNL. Reduction of selenite to elemental selenium nanoparticles by activated sludge. Environ Sci Pollut Res 2016; 23:1193–1202.

Jørgensen BB, Bak F, Jørgensen BB, Bak F. Pathways and microbiology of thiosulphate transformations and sulfate reduction in a marine sediment (Kattegat, Denmark). Appl Environ Microbiol 1991; 57:847-856.

Large PJ. Methylotrophy and methanogenesis. Van Nostrand Reinhold 1983; 53:1689-1699.

Lee WS, Chua ASM, Yeoh HK, Ngoh GC. A review of the production and applications of waste-derived volatile fatty acids. Chem Eng J 2014; 235:83–99.

Lenz M, Lens PNL. The essential toxin: The changing perception of selenium in environmental sciences. Sci Total Environ 2009; 407:3620–3633.

Letavayova L, Vlckova V, Brozmanova J. Selenium: from cancer prevention to DNA damage.

Toxicology 2006; 227:1–14.

Liamleam W, Annachhatre AP. Electron donors for biological sulfate reduction. Biotechnol Adv 2007; 25:452-463.

Lin Y, He Y, Meng Z, Yang S. Anaerobic treatment of wastewater containing methanol in upflow anaerobic sludge bed (UASB) reactor. Front Environ Sci Eng China 2008; 2:241-246.

Lu X, Zhen G, Chen M, Kubota K, Li YY. Biocatalysis conversion of methanol to methane in an upflow anaerobic sludge blanket (UASB) reactor: Long-term performance and inherent deficiencies. Bioresour Technol 2015; 198:691-700.

Luo JH, Chen H, Hu S, Cai C, Yuan Z, Guo J. Microbial selenate reduction driven by a denitrifying anaerobic methane oxidation biofilm. Environ Sci Technol 2018; 52:4006–4012.

Matanhike, L., 2017. Anaerobic biotrickling filter for the removal of volatile inorganic and organic compounds (Master's thesis). UNESCO-IHE, Delft, the Netherlands.

Mehdi Y, Hornick JL, Istasse L, Dufrasne I. Selenium in the environment, metabolism and involvement in body functions. Molecules 2013; 18:3292–3311.

Nancharaiah YV, Lens PNL. Ecology and biotechnology of selenium-respiring bacteria. Microbiol Mol Biol Rev 2015a; 79:61-80.

Pettine M, McDonald TJ, Sohn M, Anquandah GAK, Zboril R, Sharma VK. A critical review of selenium analysis in natural water samples. Trends Environ Anal Chem 2015; 5:1–7.

Qian J, Liu R, Wei L, Lu H, Chen G. System evaluation and microbial analysis of a sulfur cycle-based wastewater treatment process for co-treatment of simple wet flue gas desulfurization wastes with freshwater sewage. Water Res 2015; 80:189-199.

Qian J, Wang L, Wu Y, Bond PL, Zhang Y, Chang X, et al. Free sulfurous acid (FSA) inhibition of biological thiosulfate reduction (BTR) in the sulfur cycle-driven wastewater treatment process. Chemosphere 2017; 176:212-220.

Shen Y, Buick R. The antiquity of microbial sulfate reduction. Earth Sci Rev 2004; 64:243-272.

Speece RE. Anaerobic biotechnology for industrial wastewater treatment. Environ Sci Technol

1983; 17:416A–427A.

Tan LC, Espinosa-Ortiz EJ, Nancharaiah YV, van Hullebusch ED, Gerlach R, Lens PNL. Selenate removal in biofilm systems: effect of nitrate and sulfate on selenium removal efficiency, biofilm structure and microbial community. J Chem Technol Biotechnol 2018; 93:2380-2389.

Tan LC, Nancharaiah YV, van Hullebusch ED, Lens PNL. Selenium: environmental significance, pollution, and biological treatment technologies. Biotechnol Adv 2016; 34:886-907.

TRI Program. 2017 TRI Factsheet for Chemical METHANOL. 2017.

Van den Brand TPH, Roest KK, Chen GH, van Brdjanovic DD, Loosdrecht MCM. Occurrence and activity of sulphate reducing bacteria in aerobic activated sludge systems, World J Microbiol Biotechnol 2015: 31:507–516.

Weijma J, Chi TM, Hulshoff Pol LW, Stams AJM, Lettinga G. The effect of sulphate on methanol conversion in mesophilic upflow anaerobic sludge bed reactors. Process Biochem 2003; 38:1259-1266.

CHAPTER 4

Performance of a biotrickling filter for anaerobic utilization of gas-phase methanol coupled to thiosulphate reduction and resource recovery through volatile fatty acids production

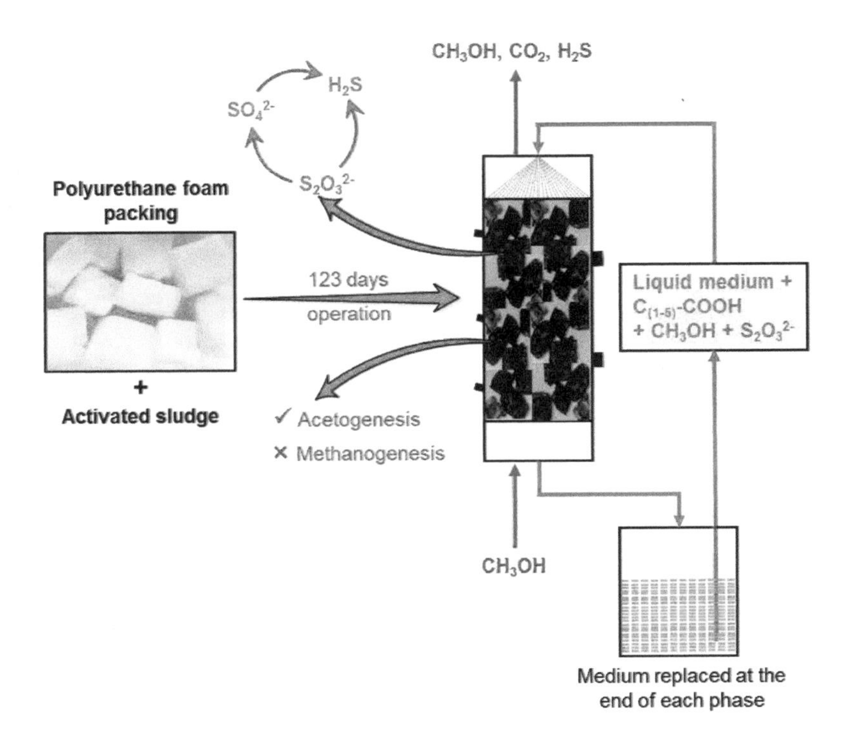

This chapter has been modified and published as:

Eregowda, T., Matanhike, L., Rene, E.R., Lens, P.N.L., 2018. Performance of a biotrickling filter for anaerobic utilization of gas-phase methanol coupled to thiosulphate reduction and resource recovery through volatile fatty acids production. Bioresour. Technol. 263, 591–600.

Abstract

The anaerobic removal of continuously fed gas-phase methanol (2.5-30 g/m^3.h) and the reduction of step-fed thiosulphate (1000 mg/L) was investigated in a biotrickling filter (BTF) operated for 123 d at an empty bed residence time (EBRT) of 4.6 and 2.3 min. The BTF performance during steady step feed and special operational phases like intermittent liquid trickling in 6 and 24 h cycles and operation without pH regulation were evaluated. Performance of the BTF was not affected and nearly 100% removal of gas-phase methanol was achieved with an EC_{max} of 21 g/m^3.h. Besides, > 99% thiosulphate reduction was achieved, in all the phases of operation. The production of sulphate, H_2S and volatile fatty acids (VFA) was monitored and a maximum of 2500 mg/L of acetate, 200 mg/L of propionate, 150 mg/L of isovalerate, 100 mg/L isobutyrate was produced.

Keywords: Biotrickling filter (BTF); gas-phase methanol; volatile fatty acid; anaerobic; thiosulphate reduction; steady and intermittent BTF operation.

4.1. Introduction

Among the primary energy consumers in the industrial sector, the paper industry is the fourth largest with an estimate of 8 exajoules (EJ), wherein 2.3 EJ is from the black liquor and wood waste (Bajpai, 2016). Black liquor from pulp washing (dissolved solids content of 14-18%) is usually concentrated in a multi-effect evaporation plant and the condensates from the black liquor digesters and evaporators typically contain 250-12000 mg/L and 375-2500 mg/L of methanol, respectively (Meyer and Edwards, 2014b; Rintala and Puhakka, 1994). As a means to save energy, reduce fresh water intake and organic load to the wastewater treatment plant, the condensate is subjected to a stripping process. During the stripping process, methanol and other volatile organic compounds (VOC) from the condensate accumulate in overhead vapour as non-condensable gases (NCG) and are generally incinerated (Siddiqui and Ziauddin, 2011; Suhr et al., 2015). Considering the nature of the emissions resulting from the incineration of NCG, biotrickling filters (BTF) have been regarded as an efficient, cost-effective, alternative biological cleaner production process for the treatment of waste gases containing methanol (Jin et al., 2007; ÓscarJ. J. Prado et al., 2004; Prado et al., 2006; Ramirez et al., 2009; Rene et al., 2010).

Most of the gas-phase methanol degradation studies have been performed under aerobic conditions using pure or mixed microbial cultures with the final oxidation product of gas-phase methanol degradation being CO_2. No previous reports have focused on the anaerobic treatment of methanol-rich NCG, additionally utilizing methanol from these NCG for production of economically viable products like volatile fatty acids. Furthermore, the usage of white wastewater from the pulp and paper industries that contain high concentrations (0.5-4.5 g/L) (Pokhrel and Viraraghavan, 2004) of oxidised forms of sulphur like sulphate (SO_4^{2-}), thiosulphate ($S_2O_3^{2-}$), sulphite (SO_3^{2-}) as the trickling medium to bring about the reduction of sulphur oxyanions has not been explored.

Recent studies have reported volatile fatty acids (VFA) as potential renewable carbon source due to their widespread industrial application. Jones et al. (2015) term acetate as *'one of the world's most important chemicals'* considering its application in the food, pharmaceutical and polymer industry and as a precursor to fuels and chemicals (Choudhari et al., 2015; Jones et al., 2017, 2015). VFA synthesis and recovery from sewage sludge, food and fermentation waste is a potential approach for chemical oxygen demand (COD) removal and carbon capture from the waste streams (Longo et al., 2014; Singhania et al., 2013; Zhou et al., 2017). Furthermore, microbial electrosynthesis and other bioelectrochemical systems are being studied for resource recovery through the production of VFA from high COD waste streams and reuse of CO_2 (Sadhukhan et al., 2016; Shemfe et al., 2018).

Industrial scale BTF operation require a continuous power supply for feed, i.e. the waste gas and nutrient medium circulation (Brinkmann et al., 2016). Deliberate shut down of liquid circulation during night time, holidays and weekends are common operational scenarios in the industries. Intermittent nutrient medium circulation is considered as an effective strategy to avoid filter bed clogging resulting from excessive biomass growth in aerobic BTF (Cox and Deshusses, 2002; Kim et al., 2005). Rene et al. (2018) recommended the simulation of intermittent feeding patterns in lab-scale studies as a pre-requisite to understand the transient performance of the bioprocess that could prevail in industrial situations. Several studies have reported BTF operation with intermittent supply of the pollutant and liquid media circulation for treatment of gaseous pollutants such as a mixture of oxygenated volatile organic compounds (ethanol, ethyl acetate, and methyl-ethyl ketone) (Sempere et al., 2008), toluene and xylene mixture (Rene et al., 2018), methyl acrylate (H. Wu et al., 2016) and BTEX (benzene, toluene, ethylbenzene and xylene) (Padhi and Gokhale, 2017).
In this study, the anaerobic utilization of gas-phase methanol (2.5-30 $g/m^3.h$) was tested in a BTF packed with a mixture of 50% polyurethane foam and 50% pall rings along with the use of $S_2O_3^{2-}$ (~ 1000 mg/L) containing synthetic wastewater as the trickling liquid. The BTF was operated continuously for 123 d in order to test the influence of empty bed residence time (EBRT: 4.6, 2.3 min), intermittent liquid trickling with 6 h and 24 h of wet and dry periods and operation without pH regulation on the elimination capacity (EC) and removal efficiency (RE) of the BTF in terms of gas-phase methanol removal, $S_2O_3^{2-}$ reduction rate and production of different VFA.

4.2. Materials and methods

4.2.1. Source of biomass and medium composition

The BTF was inoculated with anoxically maintained activated sludge (2.86 ± 0.02 g/L volatile suspended solids) collected from the Harnaschpolder wastewater treatment plant (Delft, the Netherlands). The characteristics of the sludge in terms of alkali, alkaline earth and heavy metals were as described in Matanhike (2017). The sludge consisted of 69 mg/L sulfate and in terms of trace elements, the sludge comprised of (in μg/L) Sodium (557), Magnesium (263),

Aluminium (179), Potassium (352), Calcium (147), Manganese (6,5), Iron (296) Copper (7.7) and Nickel (1.1).

The seeding was done anoxically (nitrogen influx of 0.96 m^3/d) by recirculating 4 L of activated sludge for 4 d. The solution containing the sludge was replaced with growth medium and the filter bed was acclimatized to methylotrophy by feeding gas-phase methanol (0.3 ± 0.1 g/m^3) for a duration of 7 d. DSM 63 media modified to eliminate or replace sulfate salts with chlorides was used as the nutrient growth medium and consisted of 0.5 g K_2HPO_4, 1.0 g NH_4Cl, 0.1 g $CaCl_2 \cdot 2H_2O$, 1.0 g $MgCl_2 \cdot 6H_2O$, 1 g yeast extract and 0.5 mg resazurin as redox indicator per 980 mL of MilliQ water, autoclaved at 121°C for 20 min, followed by the addition of 10 mL each of 10 g/L of sodium thioglycolate solution and 10 g/L of ascorbic acid solution. The initial pH of the growth medium was set at 7.5 (± 0.1) (DSM 63).

4.2.2. Biotrickling filter (BTF)

The schematic of the laboratory scale BTF is shown in ***Figure 4.1***. It consisted of a filter bed placed in an airtight plexiglass cylindrical column of 12 cm inner diameter and 50 cm packing height, corresponding to a working volume of 4.6 L, provided with sampling ports for collecting the gas and liquid samples. The BTF was packed with a mixture of 50% polyurethane foam (98% porosity and a density of 28 kg/m^3) cut into cubes of 1-1.5 cm^3 and 50% plastic pall rings (specific surface area of 188 m^2/m^3 and bulk density of 141 kg/m^3). The filter bed was supported between two perforated plates placed at the top and bottom of the BTF. The liquid medium trickled on the filter bed from the top at a rate of 360 L/d, corresponding to a residence time of 18.4 min. The liquid drain was collected into an air tight medium collection tank (3 L) equipped with a pH probe (Sen-Tix WTW, the Netherlands) and pH meter (pH/mV transmitter, DO9785T Delta OHM, the Netherlands). 2.5 L of growth medium was recirculated to the filter bed with the help of a peristaltic pump (Masterflex L/S Easy-Load II, Metrohm, the Netherlands).

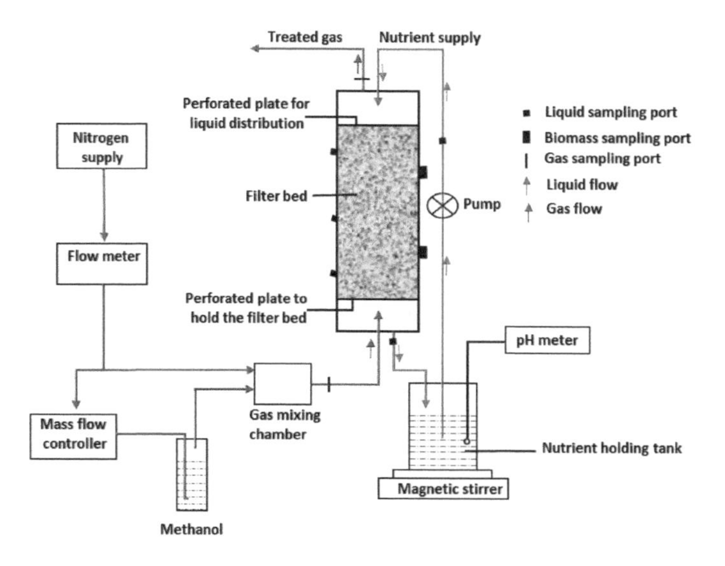

Figure 4.1: *Schematic of the biotrickling filter (BTF). Red arrows indicate the direction of gas flow, while blue arrows indicate the direction of liquid flow.*

Nitrogen gas was divided into a major and minor flow. The minor nitrogen flow was connected to a water lock apparatus containing 100% methanol through a mass flow controller (SLA 5850 MFC, Brooks instrument USA) which sparged the liquid methanol to the gas-phase. The desired concentration of methanol was achieved by regulating the flow rate of the mass flow controller. Nitrogen from the major flow and the gas-phase methanol were mixed in a mixing chamber and the homogeneous gas mixture was passed continuously to the BTF from the bottom at a flow rate of 1.44 m^3/d in all the experimental phases, except P4 where the gas flow rate was doubled (***Table 4.1***).

4.2.3. Operational phases of the BTF

The BTF operational period was divided into 8 phases based on the feeding pattern of liquid-phase $S_2O_3^{2-}$ (***Table 4.1***). The gas-phase inlet concentration of methanol was gradually increased from 0.2 to 1.15 g/m^3 (***Figure 4.2a***). The empty bed residence time (EBRT) was 4.6 min in all the phases, except phase P4 where the effect of gas-phase methanol shock load was studied by lowering the EBRT to 2.3 min (Eq. 4.1). Phase P5 was a recovery phase from the methanol shock load. Phases P6-P8 were special operations with respect to variations in the liquid medium. In phase P6, the BTF was subjected to 6 h cycles of intermittent nutrient medium trickling by switching the liquid medium between continuous recirculation and no recirculation. In phase P8, the duration of the intermittent trickling cycle was increased from 6 h to 24 h. The pH of the liquid medium was manually maintained in the range of 7.5-8.0 in all the phases, except P7, through dropwise addition of 4 M HCL when it increased > 8.0. In phase P7, the pH was not regulated and the variation of medium pH as a function of $S_2O_3^{2-}$ reduction and VFA production was studied. Phases P1, P2, P3, P6, and P8 consisted of 2 cycles each of $S_2O_3^{2-}$ feed under similar conditions. The recirculating liquid medium was changed after each cycle except in phases P1, P2 and P8.

4.2.4. Performance evaluation of the BTF

The parameters for BTF operation and performance evaluation are as follows:

Empty bed residence time (min)

$$EBRT = \frac{V}{Q} \qquad \text{Eq. 4.1}$$

Inlet loading rate (ILR, g/m^3.h)

$$ILR = \frac{Q \times C_{meth\text{-}in}}{V} \qquad \text{Eq. 4.2}$$

Elimination capacity (EC, g/m^3.h)

$$EC = Q \times \frac{C_{meth\text{-}in} - C_{meth\text{-}out}}{V} \qquad \text{Eq. 4.3}$$

Removal efficiency (RE, %)

$$RE = \frac{(C_{meth\text{-}in} - C_{meth\text{-}out})}{C_{meth\text{-}in}} \times 100 \qquad \text{Eq. 4.4}$$

Methanol consumed (g/d)

$$\text{Methanol consumed} = \frac{\sum_{cycle}(C_{meth\text{-}in} - C_{meth\text{-}out}) - (C_{liq\text{-}last} - C_{liq\text{-}first})}{\text{Number of days in a cycle}} \qquad \text{Eq. 4.5}$$

where V is the volume of the reactor (m^3), Q is the gas flow rate (m^3/h), $C_{meth\text{-}in}$ and $C_{meth\text{-}out}$ are the inlet and outlet gas-phase methanol concentrations in the BTF (g/m^3), respectively. $C_{(liq\text{-}last)}$ and $C_{(liq\text{-}first)}$ are the methanol concentrations in the liquid-phase on the last and the first day (mg/L), respectively. VFA production was assessed in terms of the highest concentration of individual VFA (mg/L) in the respective cycles. In phase P6, as a means to investigate and monitor the inconsistent drop in RE from 90 to 40%, the BTF was sampled and analysed at 2 h interval in the second cycle.

The thiosulphate ($S_2O_3^{2-}$) reduction rate (mg/L.d) was calculated as the concentration of $S_2O_3^{2-}$ added at the beginning of a cycle divided by the number of days required to achieve a concentration less than 10 mg/L, except in phase P2C2, where in the subsequent phase was started when the $S_2O_3^{2-}$ concentration was ~ 130 mg/L. In all the phases, the subsequent cycle of $S_2O_3^{2-}$ addition was not started immediately after complete reduction of $S_2O_3^{2-}$ in order to analyse the concentration of sulphate (SO_4^{2-}) and hydrogen sulphide (H_2S). H_2S production (g H_2S/cycle) in the gas-phase was calculated by estimating the area under the curve of an H_2S concentration profile for each cycle multiplied by the gas flow rate.

***Table 4.1**: Summary of operational parameters, $S_2O_3^{2-}$ reduction rate, sulphur recovery, methanol consumed for $S_2O_3^{2-}$ reduction and methanol recovered as VFA during different phases of BTF operation.*

Phase	Cycle	Operation time (d)	Gas flow rate (m^3/d)	EBRT (min)	pH	$S_2O_3^{2-}$ fed (g/cycle)	Time (d)[1]	$S_2O_3^{2-}$ reduction rate (mg/L.d)	H_2S produced (g/m^3.cycle)	S recovery (%)[2]	Methanol consumed (Eq. 4.5) (g/cycle)	Methanol consumed by SRBs (mg/cycle)	Methanol recovered as VFA (g/cycle)	C recovery as VFA (%)[4]
P1	C1	0-15	1.44	4.6	7.5-8.0	1.50	6	100	0.66	105.1	7.23	0.43	Δ	
	C2	15-22	1.44	4.6	7.5-8.0	2.04	3.5	233	0.49	57.3	3.37	0.58	4.9	46.2
P2	C1	22-29	1.44	4.6	7.5-8.0	2.22	5.5	162	0.97	103.7	9.01	0.64	Δ	
	C2	29-41	1.44	4.6	7.5-8.0	2.58	5	207	1.23	113.1	14.18	0.74	4.7	20.3
P3	C1	41-56	1.44	4.6	7.5-8.0	2.35	2.5	376	0.97	98.2	20.42	0.67	3.2	15.7
	C2	56-68.5	1.44	4.6	7.5-8.0	2.46	3.5	281	0.46	44.2	18.74	0.70	1.3	7
P4	C1	68.5-75.5	2.88	2.3	7.5-8.0	2.53	6	168	2.43	227.9	19.25	0.72	1.95	10.12
P5	C1	75.5-83	1.44	4.6	7.5-8.0	2.44	7.5	130	1.07	103.7	12.73	0.70	1.12	8.8
P6	C1	83-92.15	1.44	4.6	7.5-8.0	2.48	8	124	1.07	102.3	10.95	0.71	0.9	8.2
	C2	92.15-100	1.44	4.6	7.5-8.0	2.25	7.8	115	0.84	88.9	14.81	0.64	1.67	11.3
P7	C1	100-111.5	1.44	4.6	*	2.38	7.5	127	1.52	151.4	18.55	0.68	1.32	7.1
P8	C1	111.5-119	1.44	4.6	7.5-8.0	2.35	5	188	0.94	95.5	14.04	0.67	Δ	
	C2	119-123	1.44	4.6	7.5-8.0	2.45	4	245	0.38	37.1	12.04	0.70	0.93	3.6
						Total = 30.03 g			Total = 18.8 g	Average = 103%	Total = 175.4 g	Average = 8.58 g	Total = 22 g	Average = 12.54%

*pH was not regulated.

[1]Time required for $S_2O_3^{2-}$ concentration to reach values < 10 mg/L, except for phase P6C2 where the concentration was ~ 130 mg/L.

[2]Stoichiometric fraction of H_2S recovered in the gas-phase per unit of $S_2O_3^{2-}$ fed after subtracting the theoretical H_2S contribution from thioglycolate used in the media.

[Δ]Liquid medium was not changed.

[3]Methanol used by SRBs for $S_2O_3^{2-}$ reduction, as per the equation: $CH_3OH + S_2O_3^{2-} + H^+ \rightarrow HCO_3^- + 2HS^- + H_2O$

[4]Stoichiometric fraction of methanol recovered as VFA. VFA recovery was calculated as the difference between the final and initial concentration

4.2.5. Analytical methods

The parameters monitored in this study were $S_2O_3^{2-}$, SO_4^{2-}, H_2S, gas and liquid-phase methanol, methane and VFA. Liquid samples were filtered using 0.2 μm syringe filters and stored at 4 °C before analysis. All analysis was carried out in duplicates and the data presented in the plots are the average value with a standard deviation of < 1.0%. $S_2O_3^{2-}$ and SO_4^{2-} concentration were analysed using an ion chromatograph (ICS-1000, IC, Dionex with AS-DV sampler) as described by Ababneh et al. (2011). Methane was measured using a gas chromatograph (GC 3800, VARIAN, the Netherlands) according to the procedure described by Cassarini et al. (2017). The VFA concentration was analysed using a gas chromatograph (Varian 430-GC, Varian Inc., USA), equipped with a CP WAX-58 CB column and a flame ionization detector (FID) with helium as carrier gas. The liquid-phase and the inlet and outlet gas-phase methanol concentration were analysed by gas chromatography (Bruker Scion 456-GC, SCION Instruments, the Netherlands), using helium as the carrier gas, equipped with an FID (250 °C), Restek stabilwax column (30 m × 0.32 mm × 0.1 μm), injector (200 °C) and oven (60-250 °C). H_2S in the gas-phase was measured using a portable multi-gas sensor (Dräger X-am 7000, Dräger, Germany).

4.3. Results and discussion

4.3.1. Methanol utilization

The gas-phase methanol fed to the BTF, at loading rates varying between 2.5-30 $g/m^3.h$, was the sole electron donor and carbon source available for the biomass in the BTF. The methanol inlet, outlet gas-phase concentration, ILR (Eq. 4.2), EC (Eq. 4.3), RE (Eq. 4.4) and accumulation in the liquid-phase are shown in ***Figure 4.2b***. A gas-phase removal > 99% was achieved after 7 d of operation in phase P1 and in phase P2 when the ILR ranged from 2-8.5 $g/m^3.h$. In phase P3, with an increase in the ILR from 9 to 15 $g/m^3.h$, the RE gradually decreased to ~ 60% and further increased to 80% by the end of the phase. In phase P4, when the EBRT was reduced from 4.6 to 2.3 min, the ILR increased to ~ 20-30 $g/m^3.h$. Consequently, the RE dropped to 65% and thereafter it gradually improved to 71%. In phases P5, P6 and P7, the ILR was maintained at ~ 10 $g/m^3.h$ and in phase P5, on restoring the EBRT to 4.6 min, the gas-phase methanol removal rapidly improved to values > 95%. The removal of gas-phase methanol mainly occurred due to the dissolution of hydrophilic gas-phase methanol in the liquid-phase followed by anaerobic utilization by the attached biomass.

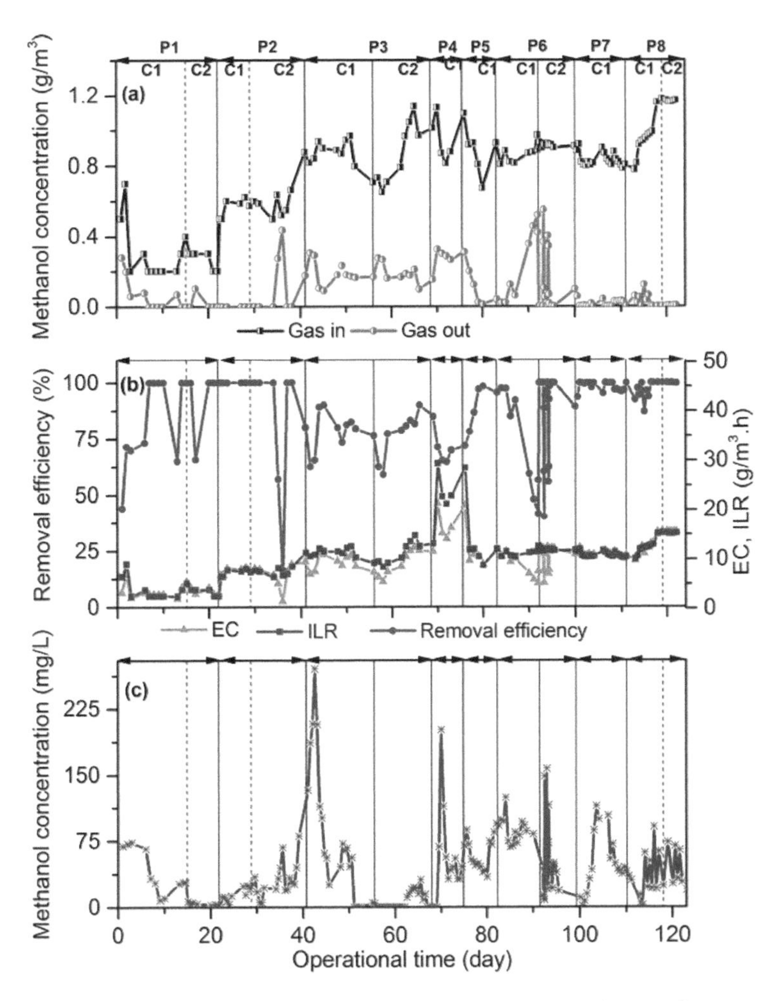

Figure 4.2: *Methanol removal in the BTF during different phases of operation: (a) inlet and outlet concentration of gas-phase methanol, (b) gas-phase removal efficiency (RE), elimination capacity (EC) and inlet loading rate (ILR) for gas-phase methanol, and (c) concentration of methanol accumulated in the liquid-phase. Full lines and dotted lines respectively indicate the replacement and no replacement of the trickling liquid.*

In phase P6, with an intermittent liquid trickling of 6 h, the RE dropped to 40% in phase P6C1 (***Figure 4.2***). To investigate the unexpected drop in RE, sampling was carried out once every 2 h in the second cycle which revealed that the RE was 100% in the dry periods. A clear trend of the drop in RE to 32% and 50%, at the beginning of the 2nd and 3rd wet period, respectively, was followed by a rapid recovery (> 90%) at the end of the cycles. The methanol molecules in the gas-phase got trapped in the water saturated voids of the filter bed during the dry cycle and when the recirculation of liquid medium was restored, the non-utilized methanol dissolved in the liquid medium. The counter-current nitrogen flow had a stripping effect on methanol that was dissolved in the liquid-phase, allowing it to escape through the BTF outlet. However, a steady RE was restored within 3 to 4 h when the dissolved methanol was stripped.

In phase P7, the unregulated pH did not affect the gas-phase methanol removal and the RE values were > 90%. In phase P8, the ILR was increased from 10 to 15 g/m^3.h. With 24 h cycles of intermittent liquid trickling, the RE varied slightly between the dry and wet periods, but the overall RE was > 80% and was nearly 100% in C2. Similar to the observation in phase P6C2, stripping of dissolved methanol from the liquid interphase occurred at the beginning of each wet periods.

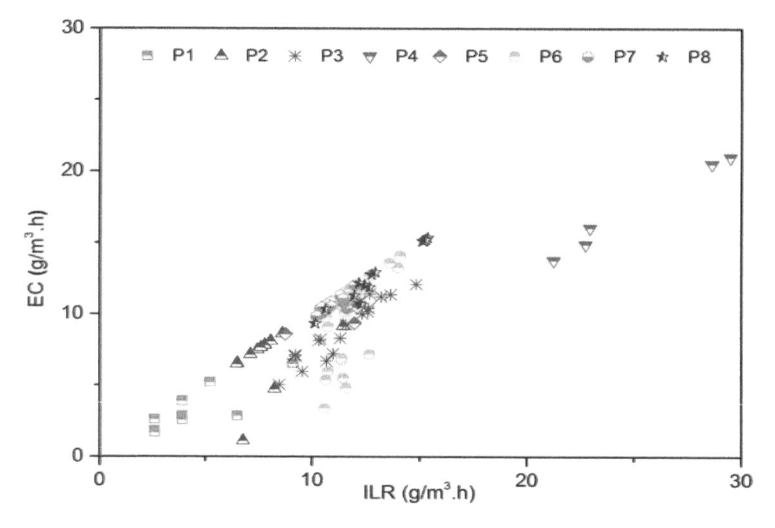

Figure 4.3: EC as a function of the ILR of gas-phase methanol during different phases of BTF operation (P1-P8).

The EC increased linearly with an increase in the ILR in phases P1 and P2. A further increase in the ILR from 10 to 15 g/m^3.h in phase P3C2 did not affect the EC of the BTF, clearly indicating that the active biomass in the filter bed was able to assimilate the increased methanol load. In phase P6, the EC varied between 3.4 and 14 g/m^3.h. EC was usually low during the beginning of a wet period, and gradually increased to values close to the ILR during the dry periods. Overall, a maximum EC (EC_{max}) of 21 g/m^3.h was achieved in phase P4 at an EBRT of 2.3 min, although the RE was 75% ***(Figure 4.3)***.

Gas-phase methanol removal in BTF has been studied extensively under aerobic conditions and EC_{max} values of 2160 $g/m^3.h$ (Ramirez et al., 2009), 552 $g/m^3.h$ (Prado et al., 2006), 6.4 $g/m^3.h$ (Jin et al., 2007) and 175 $g/m^3.h$ (ÓscarJ. J. Prado et al., 2004) have been reported at an EBRT of 20 to 265 s (Ramirez et al., 2009), 47 s and 70 s (Prado et al., 2006), 17 s (Jin et al., 2007), 83.4 s and 146.4 s (Rene et al., 2010), 28 s, 24 s, 45 s, 60 s, and 120 s (Palomo-Briones et al., 2015). Even though the EC_{max} of 21 $g/m^3.h$ reported in this study is much lower and the EBRT of 2.3 min is much higher in comparision, it is noteworthy to mention that the biomass growth rate in anaerobic process is much lower than in the aerobic process and the EC values are expected to be low. The RE was 100% during the last phase of BTF operation implying that the biomass was not inhibited by the methanol load. BTF with intermittent medium trickling under aerobic condition has been studied for oxygenated VOC and methyl acrylate removal (Sempere et al., 2008; H. Wu et al., 2016). Sempere et al. (2008) showed that intermittent liquid trickling is indeed beneficial to improve the RE of the gas-phase pollutant. Besides, intermittent pollutant feeding, as well as liquid trickling have been proposed as feasible strategies to prevent excess biomass growth in gas-phase BTF. No study has reported the anaerobic removal of gas-phase methanol and the effect of intermittent liquid medium trickling in a BTF. Although the problem of excess biomass growth is not a major operational concern for anaerobic systems, the intermittent trickling of liquid medium did not negatively affect the RE.

Several previous studies have reported the anaerobic mineralization of methanol in an upflow anaerobic sludge bed (UASB) reactor inoculated with anaerobic granular sludge and it has been elucidated that in a methanol and sulphate fed UASB reactor, methanol is utilized by sulphate reducing bacteria (SRB) for the reduction of SO_4^{2-} to H_2S, acetogens for production of acetate, other VFA and CO_2, as well as methanogens for the production of methane and CO_2 (Badshah et al., 2012; Paula L. Paulo et al., 2004; Weijma et al., 2003). In this study, methane in the gas-phase outlet was found to be negligible (< 0.2% *v/v*) during the entire 123 d of BTF operation, indicating that the methanogenic methylotrophs were not active. Thus, the methanol was mainly utilized by SRB for the reduction of $S_2O_3^{2-}$ to H_2S and by the acetogens for CO_2 and VFA production.

Due to the step-feeding pattern of $S_2O_3^{2-}$ reduction, the dynamics of the microbial activity varied throughout an operational cycle. The accumulation of methanol in the liquid-phase, occurred in all the phases of operation at varying concentration (0-238 mg/L), especially in the beginning in cycles P1C1, P3C1, P4, P5C1, P6C1, P6C2, and was eventually utilized by the biomass (***Figure 4.2c***). The acetate production rate was maximum at the beginning of the cycle, i.e. when the $S_2O_3^{2-}$ reduction rate was also maximum, indicating a syntrophic relation between the activity of SRB and acetogens. Weijma et al. (2002) suggested that acetogens can exist syntrophically with SRB since SRB provide the bicarbonate (2 moles of bicarbonate per mole of $S_2O_3^{2-}$ reduced) required for the acetogenesis. The acetate produced by acetogens could also be used by the SRB (Weijma et al., 2003) for $S_2O_3^{2-}$ reduction. Overall, 8.58 g of methanol was consumed by the SRBs for $S_2O_3^{2-}$ reduction during the 123 d of BTF operation (***Table 4.1***).

4.3.2. VFA evolution

Recently, many studies have focussed on VFA production from waste feedstock such as food waste and lignocellulosic biomass (Baumann and Westermann, 2016; Dahiya et al., 2015; Wang et al., 2014; Zhou et al., 2017) as a strategy for COD removal (Lee et al., 2014) and to foster a transition towards sustainability as they have a wide range of applications for the production of bioplastics and bioenergy. VFA production from food waste and lignocellulosic biomass are catabolic and they majorly involve hydrolysis of the complex organic substrate, followed by the acetogenic fermentation of sugars and fats under anaerobic conditions (Zhou et al., 2017) to produce acetate, propionate, butyrate, valerate and caproate. In this study, however, the VFA synthesis occurred by methanol utilization and carboxylate chain elongation.

A maximum production of 2500 mg/L acetate on day 30, 248 mg/L of propionate and 90 mg/L of butyrate on day 15, 105 mg/L of isobutyrate on days 20, 216 mg/L of isovalerate on day 39 and about 45 mg/L of valerate, respectively, was observed. Besides, a clear trend of both production and consumption of VFA, especially acetate was also noticed. Acetate accumulation occurred when the $S_2O_3^{2-}$ reduction rate was maximum in almost all the phases, suggesting that during peak activity of SRB, the activity of syntrophic acetogens for VFA production was also higher due to the availability of bicarbonate from the $S_2O_3^{2-}$ reduction (Weijma and Stams, 2001). Paulo et al. (2004) reported 2.5 times higher acetogenesis from methanol when bicarbonate was supplemented in the medium.

Production of propionate and butyrate varied between 25 and 50 mg/L in phases P2-P6. In phases P7 and P8, propionate was mostly below the detection limit, although some peak occurred occasionally. Liamleam and Annachhatre (2007) reported that propionate and butyrate are among the widely used electron donors for SRB, which explains the utilization of propionate and butyrate in the later phases of operation.

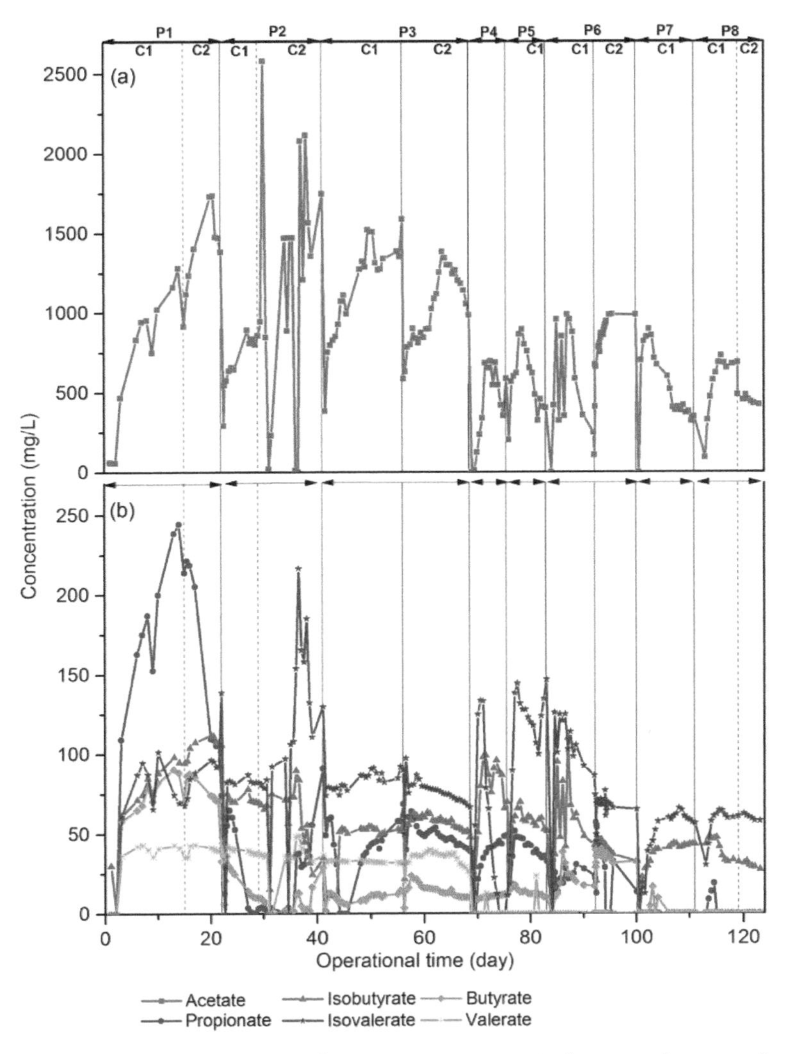

Figure 4.4: *Concentrations of acetate, propionate, isobutyrate, butyrate, isovalerate and valerate in the liquid medium during the various phases of BTF operation. The vertical lines indicate the change of cycle, the solid lines indicate the change of liquid medium and dotted lines indicate the continuation of the subsequent cycle without liquid medium replacement.*

The production of iso form of butyrate from glucose has been reported (Zhang et al., 2011) with pure culture. However no investigation has been carried out in mixed culture bioreactors. In this study, the production of isobutyrate and isovalerate occurred at higher concentrations, i.e. 110 mg/L of isobutyrate in phase P1 and 220 mg/L of isovalerate in phase P2C2, than their straight-chain counterparts. The net production of isobutyrate and isovalerate was larger in phases P4-P6 implying the stress induced by the methanol shock load and 6 h intermittent liquid medium trickling was beneficial for the production of isobutyrate and isovalerate. However, further research on the pathways involved in the anaerobic utilization of gas-phase methanol in a BTF operated with mixed cultures for the production of VFA is necessary to elucidate the exact carbon chain elongation mechanism.

During the 123 d operational period, 175.4 g of methanol was consumed by the biomass in total (***Table 4.1***). A maximum of 46.2% of carbon recovery as VFA was achieved in phase P1 and the percentage recovery gradually decreased. The recovery increased in phase P4 and P6C2, indicating that the acetogenesis was favoured by increased loading rate in P4 and the stress induced by intermittent trickling in P6C2. The actual VFA production was much higher than the final concentration as some of the VFA was utilized by the biomass after complete $S_2O_3^{2-}$ reduction (***Figure 4.4***), which is not represented in the final VFA recovery.

4.3.3. Thiosulphate ($S_2O_3^{2-}$) reduction

Sulphate reduction using different electron donors has been studied extensively (Cao et al., 2012; Liamleam and Annachhatre, 2007), whereas few studies have report $S_2O_3^{2-}$ reduction in bioreactors. The anaerobic $S_2O_3^{2-}$ reduction is generally carried out by SRB (Jørgensen et al., 1991). SRB are capable of using both methanol and acetate as energy source for the reduction of $S_2O_3^{2-}$ to HS^-, disproportionation to SO_4^{2-} and HS^-, and further reduction of SO_4^{2-} to HS^- (Cassarini et al., 2017; Weijma and Stams, 2001). Concentration profiles of $S_2O_3^{2-}$, SO_4^{2-} and H_2S during the BTF operation are shown in ***Figure 4.5***. Although the reduction rates differed, complete reduction of $S_2O_3^{2-}$ occurred in all the cycles. A maximum $S_2O_3^{2-}$ reduction rate of 376 mg/L.d was observed in phase P3C1. In phase P4, at an EBRT of 2.3 min, the $S_2O_3^{2-}$ reduction rate dropped to 168 mg/L.d, due to the reduced activity of SRB from the methanol shock load. $S_2O_3^{2-}$ reduction rate further dropped to 130 mg/L.d in the subsequent phase (P5) due to suppressed SRB activity.

During the 6 h intermittent liquid medium trickling, the $S_2O_3^{2-}$ reduction rates were 124 and 115 mg/L.d in phases P6C1 and P6C2, respectively. Track studies in phase P6C2 at 2 h intervals (***Figure 4.6***) showed that the $S_2O_3^{2-}$ reduction did not occur in the liquid medium during the dry period. Instead, the $S_2O_3^{2-}$ concentration during the 6 h of dry period increased by 17 mg/L. A similar trend was observed in the subsequent dry period of the same cycle, confirming that the active $S_2O_3^{2-}$ reduction occurred only in the filter bed. Therefore, the decrease in $S_2O_3^{2-}$ reduction rate mainly occurred due to the unavailability of $S_2O_3^{2-}$ to the attached biomass during dry periods of the phases with intermittent liquid trickling.

In phase P7, i.e. with unregulated pH, the $S_2O_3^{2-}$ reduction rate was 127 mg/L.d. Although the $S_2O_3^{2-}$ was completely available for the biomass in the filter bed, after restoring to continuous trickling mode, the reduction rate did not increase. With progressing $S_2O_3^{2-}$ reduction, the concentration of sulphide species in the dissociated HS^- form increased in the liquid medium with an increase in pH to values > 7.5. This might have caused product inhibition and shifted the $S_2O_3^{2-}$ reduction equilibrium backwards (Moosa and Harrison, 2006), leading to a decrease in the $S_2O_3^{2-}$ reduction rate. In phase P8, with wet and dry cycles of 24 h, the $S_2O_3^{2-}$ reduction rate increased to 188 mg/L.d in phase P8C1 and 245 mg/L.d in phase P8C2, respectively. the longer feast-famine interval of $S_2O_3^{2-}$ availability had a positive effect on the SRB. The pH drop from > 7.5 to 6.5 in the preceding phase (P7) also contributed to the removal of physically and biologically accumulated sulphide species from the reactor bed.

Cassarini et al. (2017) studied $S_2O_3^{2-}$ reduction coupled to anaerobic methane oxidation in a BTF inoculated with deep-sea sediment and reported a maximum $S_2O_3^{2-}$ reduction rate of 67.3 mg/L.d. According to the authors, the lower $S_2O_3^{2-}$ reduction rate was presumably due to the slow enrichment of the biomass. In this study, the average $S_2O_3^{2-}$ reduction rate per phase (***Table 4.1***) using methanol oxidation was in the range 100 to 376 mg/L.d during different phases of BTF operation. The maximum rate was achieved during 24 h intermittent liquid trickling in phase P8, while the minimum rate was observed in phase P1. The effective $S_2O_3^{2-}$ reduction rate based on the duration of liquid trickling was infact 376 and 490 mg/L.d for the phases P8C1and P8C2, respectively. This observation clearly shows that intermittent trickling is beneficial for $S_2O_3^{2-}$ removal. So far, no literature has reported the effects of intermittent liquid trickling on the reduction of $S_2O_3^{2-}$ or SO_4^{2-} in a BTF. Therefore, further investigation with adequate negative controls is necessary to quantify the influence of the feast-famine mode of operation during the intermittent liquid trickling on the $S_2O_3^{2-}$ reduction rate. Additionally, the operation of a BTF fed with sulphur oxy-anion rich industrial effluent, in intermittent liquid trickling mode, would be interesting to interpret whether this mode of operation could be applied for wastewater from the pulp and paper or mining industry, which usually contains high concentrations of $S_2O_3^{2-}$ and SO_4^{2}.

4.3.4. Disproportionation of thiosulphate ($S_2O_3^{2-}$)

Disproportionation of $S_2O_3^{2-}$ to SO_4^{2-} and H_2S has been reported in previous studies (Cassarini et al., 2017; Suarez-Zuluaga et al., 2016). In this study, a maximum of 150 mg/L SO_4^{2-} was observed after 0.5 to 1 d of $S_2O_3^{2-}$ addition, clearly indicating the occurrence of microbial disproportionation in the BTF. On an average, ~ 175 mg $S_2O_3^{2-}$/L.cycle disproportionated to SO_4^{2-} and HS^-. The highest disproportionation occurred in phase P2C2 resulting in the production of ~ 750 mg/L of SO_4^{2-}.

Further reduction of SO_4^{2-} to H_2S was observed in all the cycles. In phase P2C2, where maximum disproportionation occurred, the release of H_2S was observed for a longer period than in other phases indicating a two-step reduction of $S_2O_3^{2-}$ ($S_2O_3^{2-}$ to SO_4^{2-} and H_2S, SO_4^{2-} to H_2S). Cypionka et al. (1998) studied the pathway for $S_2O_3^{2-}$ disproportionation and showed the formation of elemental sulphur and sulphite as intermediates prior to the reduction of $S_2O_3^{2-}$ to SO_4^{2-} and H_2S. In phase P8C1, $S_2O_3^{2-}$ was dosed before the complete reduction of SO_4^{2-} and in phase P8C2, the non-reduced SO_4^{2-} remained in the BTF for the next phase, reaching a maximum of 234 mg/L before reducing to H_2S. The changes in the liquid recirculation pattern in phases P6 and P8 did not have an effect on the disproportionation of $S_2O_3^{2-}$.

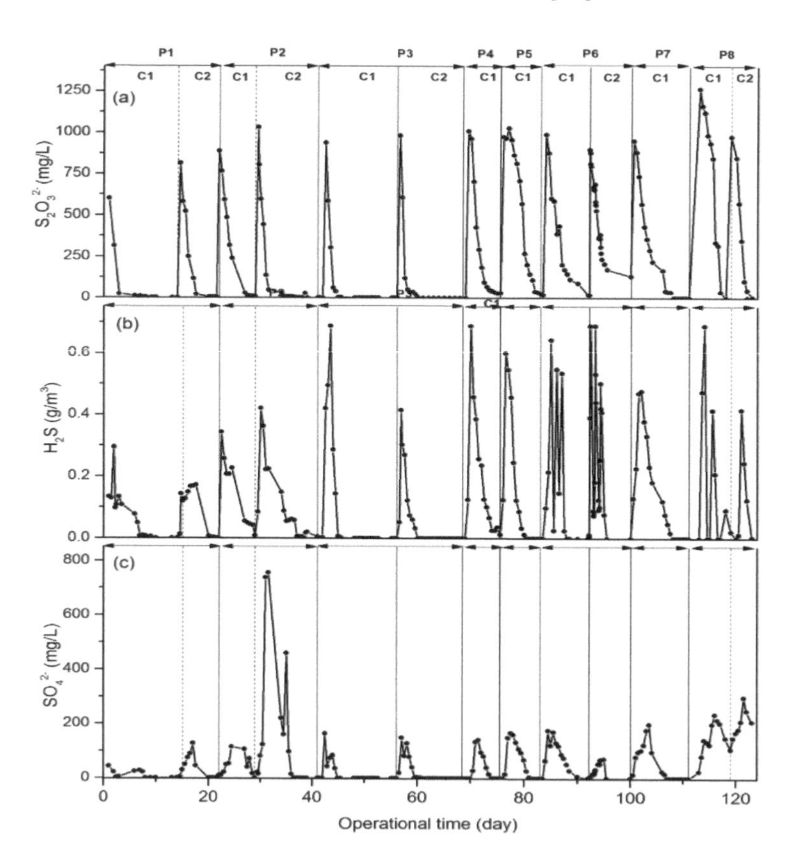

Figure 4.5: *Concentration profiles of sulphur species during different phases of BTF operation: (a) $S_2O_3^{2-}$, (b) H_2S, and (c) SO_4^{2-}.*

4.3.5. Hydrogen sulfide (H_2S) production

Theoretically, one mole of $S_2O_3^{2-}$ should be reduced to two moles of H_2S by utilizing one mole of methanol. The nutrient medium consisted of 100 mg/L of sodium thioglycolate ($C_2H_3NaO_2S$) which on complete reduction could theoretically contribute 75 mg of H_2S species per cycle of $S_2O_3^{2-}$ addition. This value was subtracted during the sulphur mass balance. A theoretical maximum recovery of H_2S was achieved in the BTF indicating the complete reduction of $S_2O_3^{2-}$ to H_2S (***Table 4.1***). The average S balance per cycle for the entire period of BTF operation was 2.2% higher than the theoretical maximum. The anomaly in the S balance between the amount of $S_2O_3^{2-}$ added and H_2S recovered could have occurred due to the fact that the S compounds existed in different physical states depending on the medium pH, biomass activity and the rate of $S_2O_3^{2-}$ reduction.

In the cycles P1C2 and P3C2, the H_2S recovery was 0.49 and 0.46 g H_2S/m^3, i.e. 57.3% and 44.2% of $S_2O_3^{2-}$ added, respectively. $S_2O_3^{2-}$ could also have accumulated as elemental sulphur and/or as polysulphides in the filter bed which was eventually reduced in the subsequent phases. In phase P6, the H_2S in the gas-phase was below detection limit during the dry periods, i.e. without liquid medium trickling. Since the gas and liquid-phase were not in contact, stripping of sulphide species from the liquid to the gas-phase did not occur, leading to a build-up of HS^- in the liquid medium. During the wet period, the H_2S concentration was as high as 0.7 g/m^3 (***Figure 4.6f***). In phase P7, wherein the pH of the liquid medium was not regulated, H_2S production was 1.52 g H_2S/m^3.cycle. This could presumably be due to the reduction of elemental sulphur and polysulphides to H_2S from the preceding phase. In phase P8, the H_2S concentration was similar to that in phase P6. H_2S is a common by-product of the anaerobic processes involving sulphur species and should be treated. Its emission in the gas-phase could be controlled by including a basic ($pH > 7$) gas trap for the effluent gas stream. The sulphides could further be used as electron donors in other processes, e.g. autotrophic denitrification (Syed et al., 2006). Furthermore, several studies have focused on chemical and biological treatment of H_2S including microbial oxidation (Syed et al., 2006) and physicochemical adsorption on bioactive material (Kanjanarong et al., 2017).

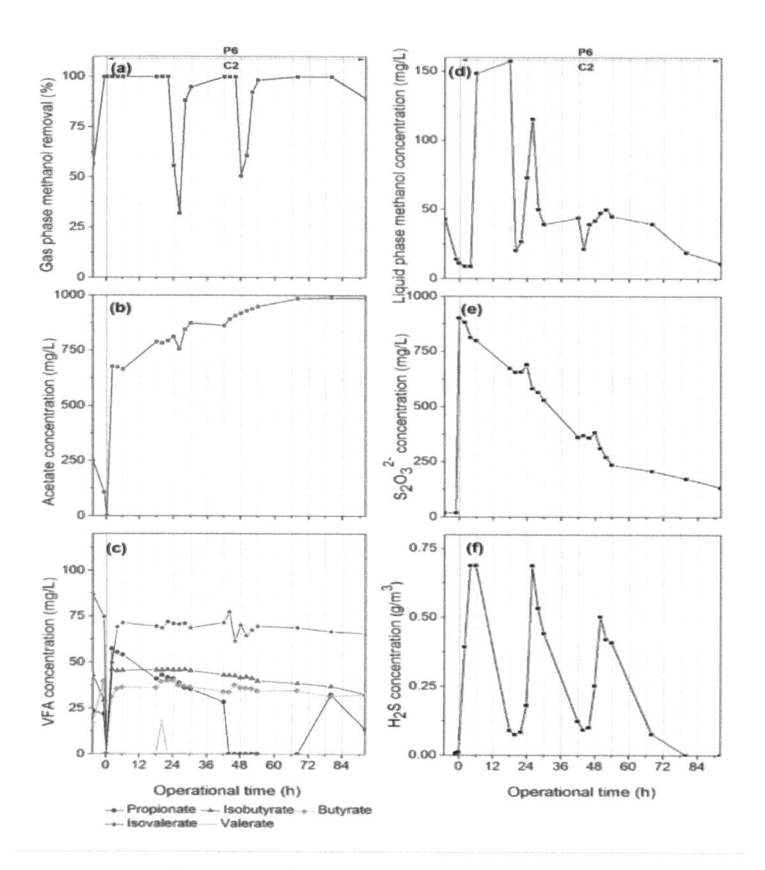

Figure 4.6: *Performance of the BTF during phase P6C2: (a) gas-phase methanol removal efficiency (RE), (b) acetate concentration in the liquid medium, (c) propionate, isobutyrate, butyrate, isovalerate and valerate concentrations in the liquid medium, (d) methanol concentration in the liquid medium, (e) $S_2O_3^{2-}$ concentration in the liquid medium, and (f) H_2S concentration in the gas-phase. The vertical grid lines indicate 6 h interval.*

4.3.6. Effect of unregulated pH

In phase P7, pH of the trickling medium was not regulated. The pH increased from 7.5 to 9.2 in 2.5 d when the $S_2O_3^{2-}$ reduction was maximum, followed by a gradual decrease and stabilization at ~ 6.6 ***(Figure 4.7)***. The trend is the result of a complex modulation of $S_2O_3^{2-}$ reduction and VFA production which are in turn affected by a pH change in the BTF. With an increase in the pH of the trickling nutrient medium from 7.5 to 9.6, the speciation of sulphide in the liquid-phase, i.e. in the form of HS^-, increased, resulting in lower concentrations of H_2S in the gas-phase. With progressive acetogenesis, the pH of the BTF gradually decreased from

9.2 to 6.6 which facilitated the formation of more H_2S (from HS^-) species in the gas-phase. During this phase, the RE of gas-phase methanol was > 99%. Besides, ~ 1000 mg/L of acetate accumulated in the liquid-phase when $S_2O_3^{2-}$ reduction was high and was eventually consumed when the SRB activity in the filter bed reduced (***Figure 4.4***).

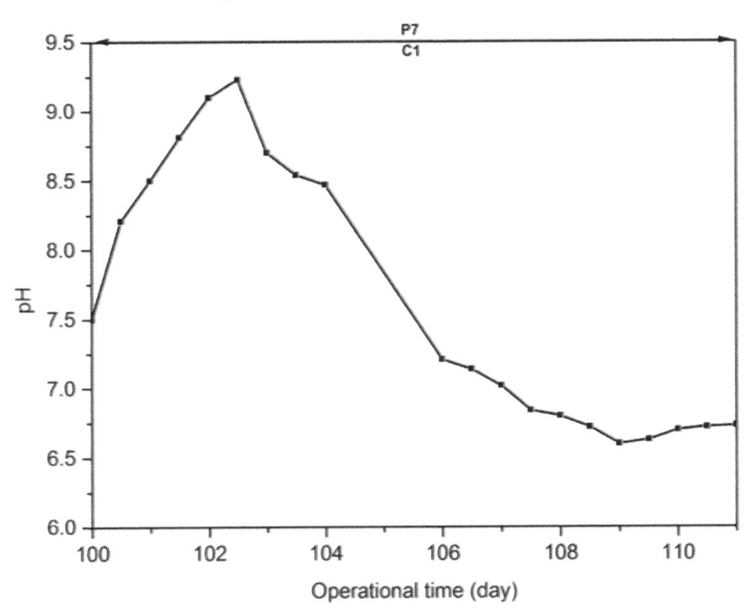

Figure 4.7: *pH profile of the liquid medium during phase P7 of BTF operation (without pH control).*

4.3.7. Practical implication and future research

The results from this study could be extended to examine the anaerobic utilization of methanol from waste gas stream like NCG in a BTF using the pulp and paper mill effluent (rich in oxidised sulphur species) as liquid medium with further possibility of resource recovery through VFA production. Further studies on gas mixtures containing methyl mercaptan (CH_3SH), dimethyl sulfide (CH_3SCH_3) and dimethyl disulfide (CH_3SSCH_3), acetone, and α-pinene which are common components of NCG from chemical pulping processes are necessary ***(Figure 4.8)***. Furthermore, the mass and energy transfer calculations and techno-chemical assessment for the complete process is interesting to further validate the practical feasibility of the process.

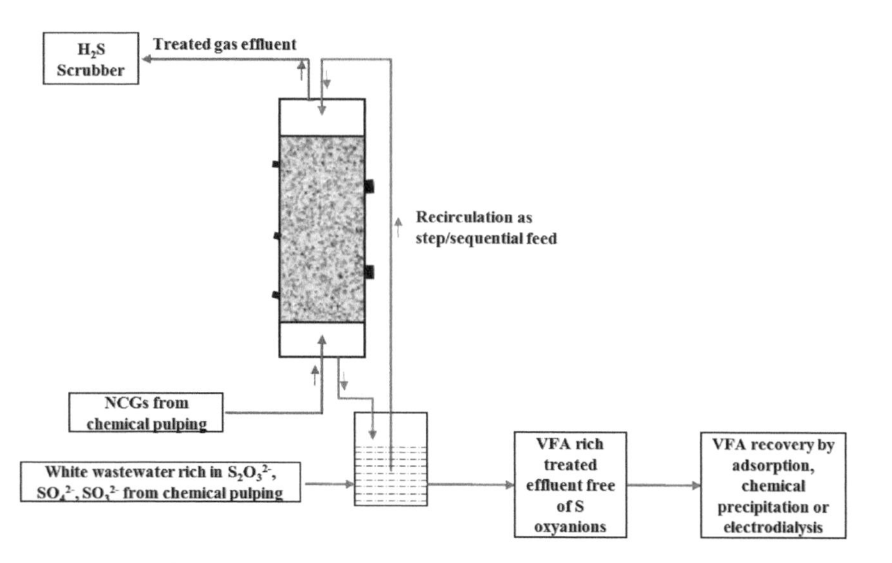

Figure 4.8: *Scheme representing the possible application of the BTF for anaerobic utilization of methanol from NCG of chemical pulping and white wastewater to achieve VFA production and reduction of S oxy-anions.*

Intermittent medium trickling and no pH control operations did not affect the RE of gas-phase methanol, but such operations affected the VFA production rate. From a VFA recovery point of view, it is evident that replacing the liquid medium at the point of complete $S_2O_3^{2-}$ reduction is beneficial in order to avoid VFA utilization by the biomass. Accordingly, VFA production from methanol could be extended for resource recovery from any methanol-rich waste stream in both gas and liquid form. The BTF operating conditions tested in this study can also be extended for the treatment of $S_2O_3^{2-}$ or SO_4^{2-} rich wastewaters since the SRB present in the biofilm facilitates the reduction process. At an industrial level, step-feeding of $S_2O_3^{2-}$ is not very common. However, in facilities having an 8-h work schedule, the production of $S_2O_3^{2-}$ or SO_4^{2-} containing wastewater stream is discontinuous and BTF performance such as the one studied, provide more insight on the effect of varying feeding patterns. Besides, it was clearly elucidated that the SRB activity can be easily restored even after long periods of $S_2O_3^{2-}$ or SO_4^{2-} unavailability. Studies on the effect of continuous $S_2O_3^{2-}$ or SO_4^{2-} feeding would be very interesting to understand the steady-state behaviour of the BTF.

More research on the activity of acetogenic biomass to optimise the VFA production and methanol removal at a higher loading rate and lower EBRT (30-60 s) would be helpful in developing high rate BTF. VFA recovery can be done in separate down-stream technologies such as electrodialysis, adsorption on ion exchange resins, fractional distillation, crystallization or precipitation (Rebecchi et al., 2016; Zhou et al., 2017). The separation of individual fatty acids from the medium is still a major challenge for full-scale systems. VFA production is presented as a novel approach for resource recovery through the anaerobic utilization of methanol-rich NCG from the pulp and paper industries.

4.4. Conclusions

The anaerobic utilization of gas-phase methanol using $S_2O_3^{2-}$ rich synthetic wastewater, was evaluated in a BTF, with the additional possibility of resource recovery through VFA production. During the 123 d of BTF operation, complete reduction of ~1000 mg/L of $S_2O_3^{2-}$ involving direct reduction, disproportionation to SO_4^{2-} and a further reduction to H_2S was observed. Acetogenesis of methanol resulted in ~2000, 250, 220 mg/L of acetate, propionate, and isovalerate, respectively. Intermittent nutrient medium trickling and unregulated pH control operations did not have a negative effect on the RE of gas-phase methanol, but it affected the VFA production rates.

Acknowledgement

The authors acknowledge the support of the lab staff from UNESCO-IHE and Harnaschpolder WWTP (Delft, the Netherlands) for providing the analytical support and activated sludge respectively. This work was supported by the Marie Skłodowska-Curie European Joint Doctorate (EJD) in Advanced Biological Waste-to-Energy Technologies (ABWET) funded from Horizon 2020 under the grant agreement no. 643071 and the Netherlands Fellowship Programmes (NFP), funded by the Netherlands government.

References

Avalos Ramirez, A., Peter Jones, J., Heitz, M., 2009. Control of methanol vapours in a biotrickling filter: Performance analysis and experimental determination of partition coefficient. Bioresour. Technol. 100, 1573-1581.

Badshah, M., Parawira, W., Mattiasson, B., 2012. Anaerobic treatment of methanol condensate from pulp mill compared with anaerobic treatment of methanol using mesophilic UASB reactors. Bioresour. Technol. 125, 318-327.

Bajpai, P., 2016. Pulp and Paper Industry: Energy Conservation. Elsevier.

Baumann, I., Westermann, P., 2016. Microbial production of short chain fatty acids from lignocellulosic biomass: current processes and market. Biomed Res. Int. 2016.

Brinkmann, T., Santonja, G.G., Yükseler, H., Roudier, S., Sancho, L.D., 2016. Best available techniques (BAT) reference document for common waste water and waste gas treatment/management systems in the chemical sector. EUR 28112 EN.

Cao, J., Zhang, G., Mao, Z., Li, Y., Fang, Z., Yang, C., 2012. Influence of electron donors on the growth and activity of sulfate-reducing bacteria. Int. J. Miner. Process. 106-109, 58-64.

Cassarini, C., Rene, E.R., Bhattarai, S., Esposito, G., Lens, P.N.L., 2017. Anaerobic oxidation of methane coupled to thiosulfate reduction in a biotrickling filter. Bioresour. Technol. 240, 214-222.

Choudhari, S.K., Cerrone, F., Woods, T., Joyce, K., O'Flaherty, V., O'Connor, K., Babu, R., 2015. Pervaporation separation of butyric acid from aqueous and anaerobic digestion (AD) solutions using PEBA based composite membranes. J. Ind. Eng. Chem. 23, 163-170.

Cox, H.H.J., Deshusses, M.A., 2002. Effect of starvation on the performance and re-acclimation of biotrickling filters for air pollution control. Environ. Sci. Technol. 36, 3069-3073.

Cypionka, H., Smock, A.M., Böttcher, M.E., 1998. A combined pathway of sulfur compound disproportionation in *Desulfovibrio desulfuricans*. FEMS Microbiol. Ecol. 166, 181-186.

Dahiya, S., Sarkar, O., Swamy, Y. V, Mohan, S.V., 2015. Acidogenic fermentation of food waste for volatile fatty acid production with co-generation of biohydrogen. Bioresour. Technol. 182, 103-113.

DSM 63, Leibniz Institut DSMZ-Deutsche Sammlung von Mikroorganismen und Zellkulturen GmbH ; Curators of the DSMZ.

Jin, Y., Veiga, M.C., Kennes, C., 2007. Co-treatment of hydrogen sulfide and methanol in a single-stage biotrickling filter under acidic conditions. Chemosphere 68, 1186-1193.

Jones, R.J., Massanet-Nicolau, J., Guwy, A., Premier, G.C., Dinsdale, R.M., Reilly, M., 2015. Removal and recovery of inhibitory volatile fatty acids from mixed acid fermentations by conventional electrodialysis. Bioresour. Technol. 189, 279-284.

Jones, R.J., Massanet-Nicolau, J., Mulder, M.J.J., Premier, G., Dinsdale, R., Guwy, A., 2017. Increased biohydrogen yields, volatile fatty acid production and substrate utilisation rates via the electrodialysis of a continually fed sucrose fermenter. Bioresour. Technol. 229, 46-52.

Jørgensen, B.B., Bak, F., 1991. Pathways and microbiology of thiosulphate transformations and sulfate reduction in a marine sediment (Kattegat, Denmark). Appl. Environ. Microbiol. 57, 847-856.

Kanjanarong, J., Giri, B.S., Jaisi, D.P., Oliveira, F.R., Boonsawang, P., Chaiprapat, S., Singh, R.S., Balakrishna, A., Kumar, S., 2017. Removal of hydrogen sulfide generated during anaerobic treatment of sulfate-laden wastewater using biochar: Evaluation of efficiency and mechanisms. Bioresour. Technol. 234, 115-121.

Kim, D., Cai, Z., Sorial, G.A., 2005. Behavior of trickle-bed air biofilter for toluene removal: Effect of non-use periods. Environ. Prog. 24, 155-161.

Lee, W.S., Chua, A.S.M., Yeoh, H.K., Ngoh, G.C., 2014. A review of the production and applications of waste-derived volatile fatty acids. Chem. Eng. J. 235, 83-99.

Liamleam, W., Annachhatre, A.P., 2007. Electron donors for biological sulfate reduction. Biotechnol. Adv. 25, 452-463.

Longo, S., Katsou, E., Malamis, S., Frison, N., Renzi, D., Fatone, F., 2015. Recovery of volatile fatty acids from fermentation of sewage sludge in municipal wastewater treatment plants. Bioresour. Technol. 175, 436-444.

Matanhike, L., 2017. Anaerobic biotrickling filter for the removal of volatile inorganic and organic compounds (Master's thesis). UNESCO-IHE, Delft, the Netherlands.

Meyer, T., Edwards, E.A., 2014. Anaerobic digestion of pulp and paper mill wastewater and sludge. Water Res. 65, 321-349.

Moosa, S., Harrison, S.T.L., 2006. Product inhibition by sulphide species on biological sulphate reduction for the treatment of acid mine drainage. Hydrometallurgy 83, 214-222.

Padhi, S.K., Gokhale, S., 2017. Treatment of gaseous volatile organic compounds using a rotating biological filter. Bioresour. Technol. 244, 270 -280.

Palomo-Briones, R., Barba De la Rosa A., Arriaga, S., 2015. Effect of operational parameters on methanol biofiltration coupled with Endochitinase 42 production. Biochem. Eng. J. 100, 9-15.

Paulo, P.L., Vallero, M.V.G., Treviño, R.H.M., Lettinga, G., Lens, P.N.L., 2004. Thermophilic (55°C) conversion of methanol in methanogenic-UASB reactors: influence of sulphate on methanol degradation and competition. J. Biotechnol. 111, 79-88.

Pelaez-samaniego, M.R., Smith, M.W., Zhao, Q., Garcia-perez, T., Frear, C., Garcia-perez, M., 2018. Charcoal from anaerobically digested dairy fiber for removal of hydrogen sulfide within biogas. Waste Management

Pokhrel, D., Viraraghavan, T., 2004. Treatment of pulp and paper mill wastewater-a review. Sci. Total Environ. 333, 37-58.

Pol, L.W.H., Lens, P.N.L., Stams, A.J.M., Lettinga, G., 1998. Anaerobic treatment of sulphate-rich wastewaters. Biodegradation 9, 213-224.

Prado, Ó.J., Veiga, M.C., Kennes, C., 2004. Biofiltration of waste gases containing a mixture of formaldehyde and methanol. Appl. Microbiol. Biotechnol. 65, 235-242.

Prado, Ó.J., Veiga, M.C., Kennes, C., 2006. Effect of key parameters on the removal of formaldehyde and methanol in gas-phase biotrickling filters. J. Hazard. Mater. 138, 543-548.

Rebecchi, S., Pinelli, D., Bertin, L., Zama, F., Fava, F., Frascari, D., 2016. Volatile fatty acids recovery from the effluent of an acidogenic digestion process fed with grape pomace by adsorption on ion exchange resins. Chem. Eng. J. 306, 629-639.

Rene, E.R., López, M.E., Veiga, M.C., Kennes, C., 2010. Steady- and transient-state operation of a two-stage bioreactor for the treatment of a gaseous mixture of hydrogen sulphide, methanol and α-pinene. J. Chem. Technol. Biotechnol. 85, 336-348.

Rene, E.R., Sergienko, N., Goswami, T., López, M.E., Kumar, G., Saratale, G.D., Venkatachalam, P., Pakshirajan, K., Swaminathan, T., 2018. Effects of concentration and gas flow rate on the removal of gas-phase toluene and xylene mixture in a compost biofilter. Bioresour. Technol. 248, 28-35.

Rintala, J.A., Puhakka, J.A., 1994. Anaerobic treatment in pulp- and paper-mill waste management: A review. Bioresour. Technol. 47, 1-18.

Sadhukhan, J., Lloyd, J.R., Scott, K., Premier, G.C., Yu, E.H., Curtis, T., Head, I.M., 2016. A critical review of integration analysis of microbial electrosynthesis (MES) systems with waste biorefineries for the production of biofuel and chemical from reuse of CO_2. Renew. Sustain. Energy Rev. 56, 116-132.

Sempere, F., Gabaldón, C., Martínez-Soria, V., Marzal, P., Penya-roja, J.M., Álvarez-Hornos, F.J., 2008. Performance evaluation of a biotrickling filter treating a mixture of oxygenated VOCs during intermittent loading. Chemosphere 73, 1533-1539.

Shemfe, M., Gadkari, S., Yu, E., Rasul, S., Scott, K., Head, I.M., 2018. Life cycle, techno-economic and dynamic simulation assessment of bioelectrochemical systems : A case of formic acid synthesis. Bioresour. Technol. 255, 39-49.

Siddiqui, N.A., Ziauddin, A., 2011. Emission of non-condensable gases from a pulp and paper mill-a case study. J. Ind. Pollut. Control 27, 93-96.

Singhania, R.R., Patel, A.K., Christophe, G., Fontanille, P., Larroche, C., 2013. Biological upgrading of volatile fatty acids, key intermediates for the valorization of biowaste through dark anaerobic fermentation. Bioresour. Technol. 145, 166-174.

Suarez-Zuluaga, D.A., Timmers, P.H.A., Plugge, C.M., Stams, A.J.M., Buisman, C.J.N., Weijma, J., 2016. Thiosulphate conversion in a methane and acetate fed membrane bioreactor. Environ. Sci. Pollut. Res. 23, 2467-2478.

Suhr, M., Klein, G., Kourti, I., Gonzalo, M.R., Santonja, G.G., Roudier, S., Sancho, L.D., 2015. Best available techniques (BAT) - reference document for the production of pulp, paper and board. Eur. Comm. 1-906.

Syed, M., Soreanu, G., Falletta, P., Béland, M., 2006. Removal of hydrogen sulfide from gas streams using biological processes - A review. Can. Biosyst. Eng. 48, 2.1-2.17

Wang, K., Yin, J., Shen, D., Li, N., 2014. Anaerobic digestion of food waste for volatile fatty acids (VFAs) production with different types of inoculum: Effect of pH. Bioresour. Technol. 161, 395-401.

Weijma, J., Chi, T.M., Hulshoff Pol, L.W., Stams, A.J.M., Lettinga, G., 2003. The effect of sulphate on methanol conversion in mesophilic upflow anaerobic sludge bed reactors. Process Biochem. 38, 1259-1266.

Weijma, J., Stams, A.J.M., 2001. Methanol conversion in high-rate anaerobic reactors, Wat Sci Tech. 8, 7-14.

Wu, H., Yin, Z., Quan, Y., Fang, Y., Yin, C., 2016. Removal of methyl acrylate by ceramic-packed biotrickling filter and their response to bacterial community. Bioresour. Technol. 209, 237-245.

Zhang, K., Woodruff, A.P., Xiong, M., Zhou, J., Dhande, Y.K., 2011. A synthetic metabolic pathway for production of the platform chemical isobutyric acid. ChemSusChem 4, 1068-1070.

Zhou, M., Yan, B., Wong, J.W.C., Zhang, Y., 2017. Enhanced volatile fatty acids production from anaerobic fermentation of food waste: A mini-review focusing on acidogenic metabolic pathways. Bioresour. Technol. 248, 68-78

CHAPTER 5

Gas-phase methanol fed anaerobic biotrickling filter for the reduction of selenate under step and continuous feeding conditions

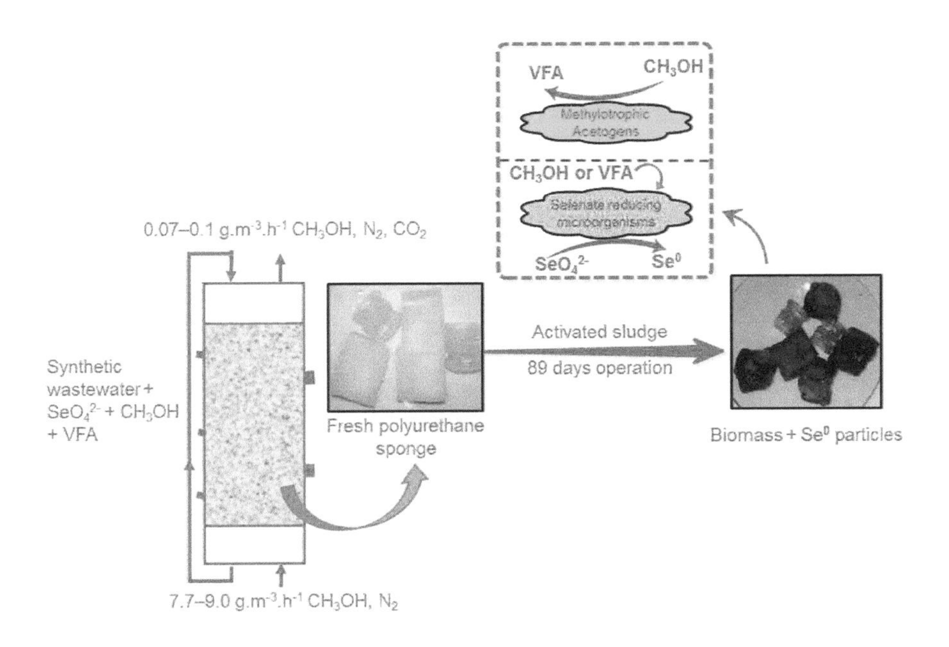

This chapter was modified and published as:

Eregowda, T., Rene, E.R., Lens, P.N.L., 2019. Gas-phase methanol fed anaerobic biotrickling filter for the reduction of selenate under step and continuous feeding conditions. Chemosphere. 225, 406-413.

Abstract

The anaerobic bioreduction of selenate, fed in step (up to 60 $mg.L^{-1}$) or continuous (~ 7 $mg.L^{-1}$) trickling mode, in the presence of gas-phase methanol (4.3–50 $g.m^{-3}.h^{-1}$) was evaluated in a biotrickling filter (BTF). During the 48 d of step-feed and 41 d of continuous-feed operations, average selenate removal efficiencies (RE) > 90% and ~ 68% was achieved, corresponding to a selenate reduction rate of, respectively, 7.3 and 4.5 $mg.L^{-1}.d^{-1}$. During the entire period of BTF operation, 65.6% of the total Se fed as SeO_4^{2-} was recovered. Concerning gas-phase methanol, the maximum elimination capacity (EC_{max}) was 46 $g.m^{-3}.h^{-1}$, with a RE > 80%. Methanol was mainly utilized for acetogenesis and converted to volatile fatty acids (VFA) in the liquid-phase.

Keywords: Gas-phase methanol, selenate reduction, biotrickling filter (BTF), step and continuous feed, volatile fatty acids (VFA)

5.1. Introduction

Selenium (Se), an element in group 16 of the periodic table, has been termed as an essential toxin for humans since it is an essential trace element at an intake of < 40 $\mu g.d^{-1}$ and it becomes toxic at an intake > 400 $\mu g.d^{-1}$ (Khoei et al., 2017). Se is widely used in the glass manufacturing, fertilizer, semiconductor and photovoltaic industries (Lenz and Lens, 2009). Se pollution in water bodies mainly occurs during flue gas desulfurization, mining operations, oil refining and agricultural runoff (C. Y. Lai et al., 2016; Tan et al., 2016). In industrial wastewaters, Se is present in the form of selenate (SeO_4^{2-}) and selenite (SeO_3^{2-}), with SeO_4^{2-} being the dominant species. For instance, in the wastewaters from flue gas desulfurization, the Se species present are SeO_4^{2-} (2778 $\mu g.L^{-1}$), SeO_3^{2-} (26.2 $\mu g.L^{-1}$) and selenocyanate (25 $\mu g.L^{-1}$) (NAMC 2010), respectively. Studies on biological Se removal processes are shifting from detoxification, i.e. by the reduction (Soda et al., 2011; Takada et al., 2008) of Se oxyanions (SeO_x^{2-}), namely SeO_4^{2-} and SeO_3^{2-}, towards the recovery of Se from bioreactors (Hageman et al., 2017; Soda et al., 2011) as quantum dots or raw material (Espinosa-Ortiz et al., 2017; Mal et al., 2017; Wadgaonkar et al., 2018). Biological processes are recommended over physico-chemical processes as the effluent SeO_x^{2-} concentrations are low and the recovery of Se is possible, along with avoiding the use of toxic chemicals (C. Y. Lai et al., 2016; Winkel et al., 2012).

Methanol is a volatile organic compound (VOC) that is commonly present in the emissions of the chemical pulping and paint industries and petroleum refineries. For example, one tonne of Kraft pulp produces 6-20 kg of CH_3OH, and 130 million tonnes of Kraft pulp are produced worldwide, corresponding to 0.8-2.6 million tonnes of CH_3OH released annually (Tran and Vakkilainnen, 2008). Biofilters (BF) and biotrickling filters (BTF) have been successfully used for the aerobic removal of volatile organic pollutants such as methanol, methyl tert-butyl ether, hexane, BTEX (benzene, toluene, ethylbenzene and xylene) and reduced sulfur compounds in gas-phase (Deshusses and Cox, 2003; Eregowda et al., 2018; Gabriel and Deshusses, 2003; Siddiqui and Ziauddin, 2011; Suhr et al., 2015). The removal of gas-phase methanol in the presence of other waste-gas components like H_2S (Jin et al. 2007), formaldehyde (Prado et al, 2006), α-pinene (Rene et al. 2010) has also been extensively tested in aerobic BF and BTF. During aerobic treatment of VOC like methanol and formaldehyde, the main end-products are CO_2, water and biomass, whereas through anaerobic treatment, recovery of useful by-products such as CH_4 and VFA is possible.

Few studies have focused on the use of gas-phase organic carbon like methane for sulfate (Bhattarai et al., 2018), SeO_4^{2-} (Lai et al., 2016) and thiosulfate (Cassarini et al., 2017) reduction and gas-phase methanol for thiosulfate reduction (Eregowda et al., 2018). Although methanol has been tested as a potential electron donor for SeO_x^{2-} reduction (Soda et al., 2011; Takada et al., 2008), methanol mineralization in the presence of SeO_x^{2-} has never been investigated in batch or continuously operated bioreactors.

The objectives of this study were, therefore, (1) to evaluate the bioreduction of SeO_4^{2-} and recovery of reduced Se from the filter bed, (2) to compare the SeO_4^{2-} reduction rates during step and continuous feeding mode of operation, and (3) to examine the gas-phase methanol utilization.

5.2. Materials and methods

5.2.1. Source of biomass and media composition

The activated sludge (2.86 ± 0.02 $g.L^{-1}$ volatile suspended solids) used as the inoculum was collected from the Harnaschpolder wastewater treatment plant (Delft, the Netherlands) which treats municipal wastewater. The characteristics of the sludge in terms of alkali, alkaline earth and heavy metals were as described in Matanhike (2017). The sludge consisted of 69 mg/L sulfate and Se was below detection limit (< 0.2 μg/L). In terms of trace elements, the sludge comprised of (in μg/L) Sodium (557), Magnesium (263), Aluminium (179), Potassium (352), Calcium (147), Manganese (6,5), Iron (296) Copper (7.7) and Nickel (1.1). The BTF was inoculated by recirculating 4 L of settled activated sludge continuously over the filter bed under nitrogen (N_2) sparging (0.96 $m^3.d^{-1}$) conditions for 4 days, following which the left over liquid-phase was replaced with 2.5 L of the trickling liquid, i.e. synthetic wastewater. DSM 63 media modified to eliminate or replace sulfate salts with chlorides was used as the trickling liquid was as follows: 0.5 g K_2HPO_4, 1.0 g NH_4Cl, 0.1 g $CaCl_2{\cdot}2H_2O$, 1.0 g $MgCl_2{\cdot}6H_2O$, 1 g yeast extract and 0.5 mg resazurin (redox indicator) per 980 mL of MilliQ water. The medium was autoclaved at 121°C for 20 min, followed by the addition of 10 mL each of 10 $g.L^{-1}$ of sodium thioglycolate solution and 10 $g.L^{-1}$ of ascorbic acid solution. The initial pH of the growth medium was adjusted to 7.5 (± 0.1). The acclimatization of the attached biomass was done by feeding 0.4 $g.m^{-3}$ of gas-phase methanol for a period of 10 d before the start of phase S1 (DSM 63).

5.2.2. Biotrickling filter set-up and operation

The BTF set-up (***Figure 5.1***) was as described previously (Eregowda et al., 2018) and consisted of an airtight plexiglass cylinder with a working volume of 4.6 L. A mixture of 50%

polyurethane foam cut into cubes of 1–2 cm^3 and 50% plastic pall rings (***Figure 5.2***) was used as the biomass support material. Polyurethane foam cubes retain the trickling liquid along with providing a high surface area (98% porosity, 2.81 $m^2.g^{-1}$ surface area and a density of 28 $kg.m^{-3}$) for biomass attachment and growth, whereas the plastic pall rings (specific surface area of 188 $m^2.m^{-3}$ and bulk density of 141 $kg.m^{-3}$) provide structural support to the filter bed along with avoiding bed compaction during long-term operation. The trickling liquid was recirculated using a peristaltic pump (Masterflex L/S Easy-Load II, Metrohm, Schiedam, the Netherlands), and trickled on top of the filter bed at a rate of 360 $L.d^{-1}$. It was collected into an airtight collection tank of 3 L, equipped with a pH meter (pH/mV transmitter, DO9785T Delta OHM, Padova, Italy). The gas flow consisted of N_2, divided into a major and minor flow, with the minor flow connected to an airtight water-lock apparatus filled with 100% methanol, through a mass flow controller (SLA 5850 MFC, Brooks instrument, Pennsylvania, USA) that sparged liquid methanol into the gas-phase. The major and minor N_2 flows were mixed in a gas mixing chamber before passing to BTF continuously in up-flow mode at a flow rate of 1.44 $m^3.d^{-1}$.

The BTF operation was carried out for 89 d in 8 phases (S1–S8) based on the SeO_4^{2-} feeding patterns as summarized in ***Table 5.1***. S1–S7 were phases with step-feeding of 50–60 $mg.L^{-1}$ of SeO_4^{2-} in the trickling liquid. Phase S8 was operated for 41 d in continuous mode wherein the SeO_4^{2-} (~ 7 $mg.L^{-1}$) rich trickling liquid was not recirculated, but fed as one-pass. The retention time (RT) was 18.4 min during the step-feed in phases S1–S7 and 22.1 h during continuous-feed in phase S8, respectively. The pH of the trickling liquid was maintained in the range of 7.5–8.0 during the entire BTF operation by the drop-wise addition of 4 M NaOH. The trickling liquid was changed at the beginning of each phase of SeO_4^{2-} addition, except in phase S2.

Table 5.1: *Overview of the BTF operational parameters.*

Phase	**Operation time (d)**	**Methanol as continuous feed ($g/m^3.h$)**	**Selenate (mg/L)**
S1	1–11	Gas-phase; 2.5–4.5	Step feed; 46
S2	11–17	Gas-phase; 0	-
S3	17–21	Gas-phase; 3.5–4.0	Step feed; 49
S4	21–31	Liquid-phase as two step feed cycles; 1 g/L	Step feed; 56
S5	31–37	Gas-phase; 0.9	Step feed; 49
S6	37–43	Gas-phase; 0.9	Step feed; 61
S7	43–49	Gas-phase; 0.9	Step feed; 50
S8	49–88	Gas-phase; 0.6 –0.7	Continuous feed; 6-8

The EBRT (Eq. 5.1) of the gas-phase methanol was 4.6 min during the entire period of BTF operation. The concentration of the gas-phase methanol fed was in the range of 0.3–4.5 g.m^{-3} in phase S1. The methanol supply was stopped in phase S2 to study the effect of microbial utilization and gas stripping (caused by the counter-current N_2 flow) on the methanol accumulated in the liquid-phase. In phase S3, the gas-phase methanol concentration was maintained in the range of 3.7–4.1 g.m^{-3} to study the effect of high gas-phase methanol loading rates on the performance of the BTF. However, phase S3 was terminated after 3 d of operation due to the accumulation of up to 5000 mg.L^{-1} methanol in the trickling liquid. In phase S4, the gas-phase methanol supply was completely stopped and the BTF was fed with 1000 mg.L^{-1} of methanol in the liquid-phase. The gas-phase methanol feed was restored in phases S5–S8 at an inlet concentration of ~ 0.7 g.m^{-3}.

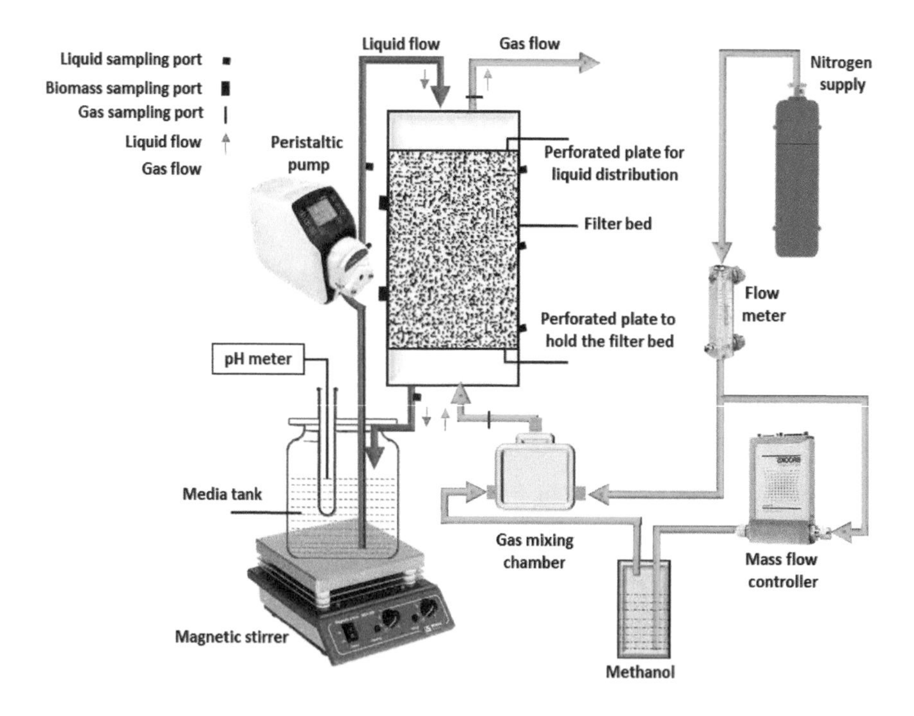

Figure 5.1: *Schematic of the biotrickling filter (BTF). Blue lines indicate the direction of liquid flow, whereas the green lines indicate the direction of the gas flow.*

To estimate the volatile organo-Se fraction, an acidic gas trap containing 100 mL of 5 M HNO_3 was included for the gas-phase effluent of the BTF in phase S8. To estimate the amount of Se in the filter bed, comprising of elemental Se (Se^0) and Se assimilated in the cell biomass after

89 d of BTF operation, randomly chosen polyurethane sponge cubes (in triplicate) were collected from the filter bed and cut into pieces of ~ 0.3 mm^3 each. The fine sponge pieces were washed with MQ water to remove the soluble forms of Se, digested overnight with an aqua regia solution (15 mL HNO_3 + 45 mL HCl) at room temperature (Gabos et al., 2014). The Se concentration of the aqua regia solution was measured using Thermo-Elemental Solaar MQZe GF95 graphite furnace atomic absorption spectroscopy (GF-AAS, PerkinElmer, Waltham, Massachusetts, USA, Se lamp at 196 nm) (Wadgaonkar et al., 2018) after suitable dilution.

5.2.3. Performance parameters of the BTF

The following parameters were considered for the evaluation of the BTF performance:

Empty bed residence time (min) $EBRT = \frac{V}{Q}$ Eq. 5.1

Inlet loading rate (ILR, $g.m^{-3}.h^{-1}$) $ILR = \frac{Q \times C_{meth\text{-}in}}{V}$ Eq. 5.2

Elimination capacity (EC, $g.m^{-1}.h^{-1}$) $EC = Q \times \frac{C_{meth\text{-}in} - C_{meth\text{-}out}}{V}$ Eq. 5.3

Gas-phase methanol removal efficiency (RE, %) $RE = \frac{(C_{meth\text{-}in} - C_{meth\text{-}out})}{C_{meth\text{-}in}} \times 100$ Eq. 5.4

Methanol consumed ($g.d^{-1}$) $= \frac{\sum_{cycle}(C_{meth\text{-}in} - C_{meth\text{-}out}) - (C_{liq\text{-}last} - C_{liq\text{-}first})}{\text{Number of days in a cycle}}$ Eq. 5.5

Retention time (RT, min) $RT = \frac{V'}{Q'}$ Eq. 5.6

Selenate reduction rate ($mg.L^{-1}.d^{-1}$) $SeRR = \frac{[SeO_4^{2-}]_i - [SeO_4^{2-}]_e}{\Delta t}$ Eq. 5.7

where, V is the volume of the reactor (m^3), Q is the gas flow rate ($m^3.h^{-1}$), $C_{meth\text{-}in}$ and $C_{meth\text{-}out}$ are, respectively, the inlet and outlet gas-phase methanol concentrations in the BTF ($g.m^{-3}$), $C_{(liq\text{-}last)}$ and $C_{(liq\text{-}first)}$ are the methanol concentrations in the liquid-phase on the last and the first day ($mg.L^{-1}$) of each phase, respectively. V' and Q' are the trickling liquid volume (L) and flow rate ($L.h^{-1}$), respectively, in the BTF.

For the phases S1-S7, $[SeO_4^{2-}]_i$ is the influent SeO_4^{2-} concentration at the beginning of a cycle and $[SeO_4^{2-}]_e$ = 0.2 $mg.L^{-1}$. Δt is the number of days required to achieve a SeO_4^{2-} concentration below 0.2 $mg.L^{-1}$ (duration of a cycle). For phase S8, $[SeO_4^{2-}]_i$ and $[SeO_4^{2-}]_e$ are the influent and effluent SeO_4^{2-} concentrations ($mg.L^{-1}$), respectively, and Δt = liquid retention time (RT).

VFA production was assessed in terms of the highest net production of individual VFA ($mg.L^{-1}$) in the trickling liquid during an operational phase.

5.2.4. Analytical methods

To evaluate the performance of the BTF, the SeO_4^{2-}, total dissolved Se, VFA, gas and liquid-phase methanol concentrations and CH_4 concentration were analysed in the influent and effluent of the BTF. The gas samples were collected in gas-tight syringes and the analysis was carried out immediately after sampling. The liquid samples were filtered using 0.45 μm syringe filters and stored at 4 °C until further analysis. Both liquid and gas sample analyses were carried out in duplicates and the data presented in the plots are the average values with a standard deviation less than 1.0%.

The concentration of SeO_4^{2-} was analysed using an ion exchange chromatograph (ICS-1000, IC, Dionex, California, USA) equipped with a AS4A 2 mm (Dionex) column at a RT of ~ 8.0 min (Mal et al., 2017). The total dissolved Se concentration was measured after centrifuging the liquid sample at 37000 × *g* (Hermle Z36 HK high-speed centrifuge, Wehingen, Germany) and analysing the supernatant for its dissolved Se content using a GF-AAS. The VFA concentration was analysed by gas chromatography (Varian 430-GC, California, USA) as described previously by Eregowda et al. (2018). The liquid-phase, as well as gas-phase inlet and outlet concentrations of methanol were analysed using a gas chromatograph (Bruker Scion 456-GC, SCION Instruments, Goes, the Netherlands) equipped with a flame ionization detector, using helium as the carrier gas (Eregowda et al., 2018). CH_4 was measured using a gas chromatograph (GC 3800, Varian, California, USA) as described previously (Cassarini et al. 2017).

5.3. Results

5.3.1. Selenate reduction

Figure 5.3 shows the SeO_4^{2-} concentration profile during BTF operation. During the first SeO_4^{2-} feeding cycle encompassing the operational phases S1 and S2, wherein the SeO_4^{2-} concentration in the influent trickling liquid was 46 $mg.L^{-1}$, a gradual SeO_4^{2-} reduction at a rate of 2.62 $mg.L^{-1}.d^{-1}$ was observed. In phase S3, although the SeRR (Eq. 5.7) was high on the 1st day, the SeO_4^{2-} concentration did not decrease further. Thus, the subsequent phase S4 was started after replacing the trickling liquid. Methanol (1000 $mg.L^{-1}$) was supplied to the BTF in the liquid-phase by adding it to the SeO_4^{2-} rich (56 $mg.L^{-1}$) trickling liquid. In phase S4, the

SeO_4^{2-} reduction efficiency was > 99% within 6.5 d. SeO_4^{2-} concentrations of 49, 61 and 50 $mg.L^{-1}$ were step-fed, respectively, in phases S5, S6 and S7, and > 99% SeO_4^{2-} reduction was observed. In phase S8, where the synthetic wastewater containing 6.0–7.2 $mg.L^{-1}$ of SeO_4^{2-} was trickled continuously at an RT (Eq. 5.6) of 22.1 h, a steady-state removal of SeO_4^{2-} was achieved. The effluent SeO_4^{2-} concentration was ~ 4.0 $mg.L^{-1}$ during the beginning of this phase and it gradually decreased to 1.5 $mg.L^{-1}$, except for short periods of offset where the effluent SeO_4^{2-} concentration was as high as 4.0 $mg.L^{-1}$.

5.3.2. Selenium mass balance and recovery

In phase S8, Se mass balance was performed in the BTF. During the 41 days of operation, 1320 mg of SeO_4^{2-} was fed to the BTF. The unreduced SeO_4^{2-} concentration in the effluent accounted for 422 mg, implying that 68% of the SeO_4^{2-} feed was reduced in the BTF. Furthermore, the total dissolved Se concentration in the effluent was 460 mg indicating that nearly 91.7% of the dissolved Se was unreduced SeO_4^{2-}. Out of the total amount of SeO_4^{2-} reduced in phase S8 of BTF operation (~ 898 mg), 103.4 (± 12.6) µg of Se escaped in the gas-phase as volatile alkylated organo-Se, accounting to < 0.022% of the Se fed to the BTF.

During the entire 89 d of BTF operation (phase S1–S8), ~ 2014 mg of SeO_4^{2-} was fed to the BTF. Considering an average removal efficiency of 95% in all the step-feed phases S1–S7 (except in phase S3), and a 68% SeO_4^{2-} removal efficiency in phase S8, a total of 1592 mg SeO_4^{2-} was reduced in the BTF, which corresponds to ~ 878.6 mg of Se. Se entrapped in the polyurethane foam cubes amounted to ~ 376 (± 48) µg Se per cube. With a filter bed of 4.6 L comprising 50% foam cubes, and the average volume of each cube being ~ 1.5 cm^3, ~ 576 mg of Se was estimated to be entrapped in the filter bed of the BTF. This implies that 65.6% of the reduced Se was recovered in the form of entrapped Se from the filter bed. It should be noted that the Se entrapped in the biomass attached to the pall rings (50% of the volume of the filter bed) was not considered in the recovery process as the biomass growth on the pall rings was negligible compared to the amount grown on the polyurethane foam cubes (***Figure 5.2***).

5.3.3. Methanol utilization

Gas-phase methanol was fed as the sole electron donor to the BTF (***Figure 5.4***). In phase S1, when the ILR (Eq. 5.2) was maintained in the range of 4.3–6.7 $g.m^{-3}.h^{-1}$, the RE (Eq. 5.4) increased from 70 to ~ 98% within 5 d and it further decreased to ~ 75% when the ILR was increased to 32–50 $g.m^{-3}.h^{-1}$ for the following 11 d (***Figure 5.5***). This increase in ILR resulted in a methanol concentration of 2700 $mg.L^{-1}$ in the liquid-phase. To monitor the utilization of the accumulated methanol in the trickling liquid, the gas-phase methanol supply to the BTF was interrupted in phase S2. The liquid-phase methanol concentration reduced from 2700 to 770 $mg.L^{-1}$ in 6 d at a rate of 255 $g.L^{-1}.d^{-1}$. The outlet gas-phase methanol concentration was ~ 0.42 $g.m^{-3}$ at the beginning of phase S2 and it gradually decreased to ~ 0.12 $g.m^{-3}$ within 6 d. During the 6 d period, about 840 mg of methanol was stripped by the upward gas flow, while 1090 mg was utilized by the biomass.

Figure 5.2: *Polyurethane foam cubes (PU) and pall rings (PR) used as the packing material for the filter bed of the BTF on (a) day 0 and (b) day 89 of BTF operation*

In phase S3, the gas-phase methanol load was maintained in the range of 47–50 $g.m^{-3}.h^{-1}$. Although an EC (Eq. 5.3) of 37.5–42.5 $g.m^{-3}.h^{-1}$ and an RE of 78–85 % was achieved, the methanol load was much higher than the utilization capacity of the biomass in the filter bed, resulting in an accumulation of up to 5000 $mg.L^{-1}$ of methanol in the liquid-phase. Since the SeO_4^{2-} reduction and acetate production rate were affected, the trickling liquid was replaced and the subsequent operational phase (phase S4) was started. Incidentally, the gas-phase methanol supply was not possible due to technical reasons in phase S4 and 1000 $mg.L^{-1}$ of methanol was supplied to the BTF along with the trickling liquid, in 2 step-feeding cycles. The supplied methanol was removed within 4.5 d in the 1st cycle and 6 d in the 2nd cycle at an average removal rate of 173 $mg.L^{-1}.d^{-1}$. Meanwhile, the average outlet gas-phase methanol concentration was 0.15 $g.m^{-3}$, implying that out of the 5000 mg of methanol fed to the BTF, 2200 mg escaped in the gas-phase and the biomass utilized ~ 2800 mg (56%) of methanol. During phases S5, S6 and S7, a steady inlet methanol load of 9.7 $g.m^{-3}.h^{-1}$ was maintained. In these phases, the RE was in the range of 80–84%, while the EC was 7.6–8.8 g.m^{-3}.h^{-1}. A steady methanol utilization was achieved in the BTF, with an accumulation of up to 590, 275, 660 mg.L^{-1} of methanol in the liquid-phase during the phases S5, S6 and S7, respectively.

In phase S8, wherein the SeO_4^{2-} containing trickling liquid was not recirculated and the gas-phase methanol ILR was steadily maintained in the range of 7.7–9.0 g.m^{-3}.h^{-1}, an EC of 6.5–8.6 g.m^{-3}.h^{-1} was achieved. In this phase, a RE > 95% was observed for gas-phase methanol at the beginning of the phase and a steady RE in the range of 80–90% was achieved for the subsequent 8 d of BTF operation. Although the accumulation of methanol in the liquid-phase occurred in the range of 130–200 mg.L^{-1} initially (days 50–68), it was only < 25 mg.L^{-1} during days 72–76 of BTF operation. To identify the end-products of methanol utilization, the gas-phase effluent of the BTF was analysed once every 3–4 d throughout the 89 d of operation for CH_4 and its concentration was < 0.5% *v/v*. CH_4 was thus not a significant end-product of methanol biodegradation in the BTF investigated. However, the trickling liquid constituted of a considerable amount of VFA resulting from the anaerobic conversion of methanol in the BTF.

5.3.4. Volatile fatty acid production

The VFA concentrations were monitored to evaluate the methanol recovery (***Figure 5.6***). During the operational phases S1 and S2, a cumulative VFA production of up to 800 mg.L^{-1} of acetate, 117 mg.L^{-1} of propionate, 35 mg.L^{-1} of isobutyrate, 56 mg.L^{-1} of butyrate, 47 mg.L^{-1} of isovalerate and 15 mg.L^{-1} of valerate were present in the liquid-phase. The phase S3 was stopped after 3 d of operation as the acetogenesis was affected by the accumulation of methanol in the liquid-phase. In phase S4, where methanol was supplied to the BTF in the liquid-phase, up to 500 mg.L^{-1} of acetate and 90 mg.L^{-1} of propionate were produced, while the concentrations of isobutyrate, butyrate and isovalerate were similar to those observed in phases S1 and S2. However, the valerate concentration gradually dropped to below detection limits (7.5 mg.L^{-1}). The acetate (and other VFA) production rate in phase S4 was maximum at the beginning of the phase and a period of acetate (and other VFA) utilization was observed towards the end of the phase S4. Before the subsequent liquid-phase methanol feeding cycle, ~ 150 mg.L^{-1} of acetate was utilized. In phases S5, S6 and S7, where a constant methanol inlet load of 9.6–9.8 g.m^{-3}.h^{-1} was applied, a stable production of ~ 350 mg.L^{-1} of acetate, 60 mg.L^{-1} of propionate and 35–50 mg.L^{-1} of other VFA was noticed. In phase S8, when the trickling liquid (pH: 7.5) was fed continuously, the effluent pH was ~ 6.5 (***Figure 5.7***). The VFA production in the effluent was stable with concentrations of ~ 200 $mg.L^{-1}$ of acetate, and ~ 20–30 $mg.L^{-1}$ each of propionate, isobutyrate, butyrate and isovalerate.

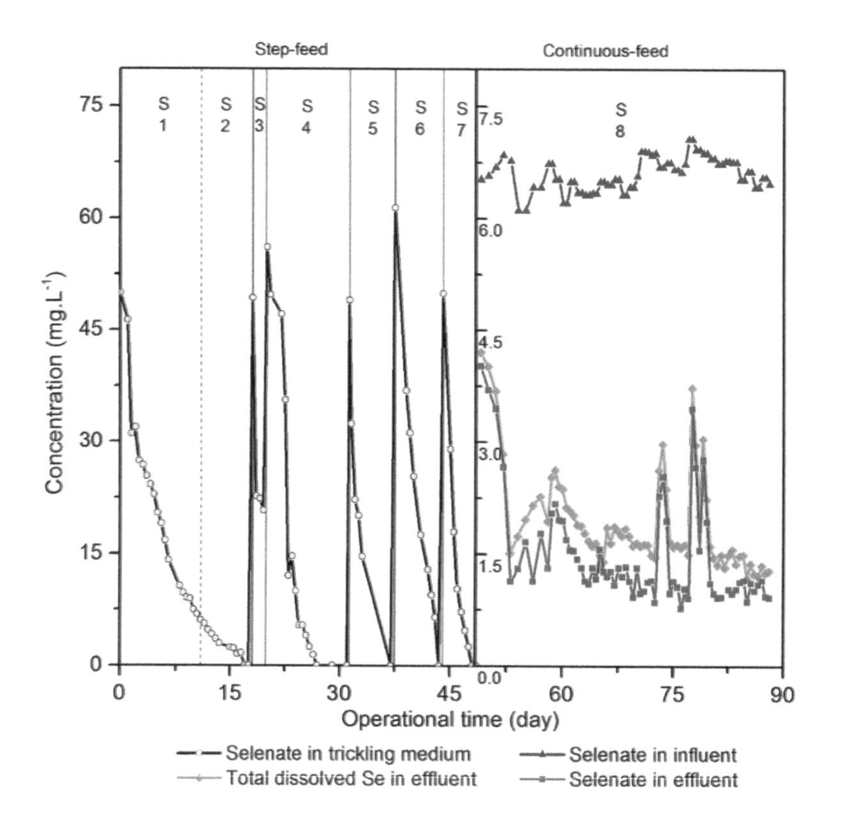

Figure 5.3: *Concentration profiles of SeO_4^{2-} during different phases of BTF operation (S1-S8) and total Se (in green) during phase S8 with continuous-feed of SeO_4^{2-}. Vertical lines indicate the change of phase, with solid lines indicating the replacement of the trickling liquid and dotted lines indicating the continuation of the subsequent phase without changing the trickling liquid.*

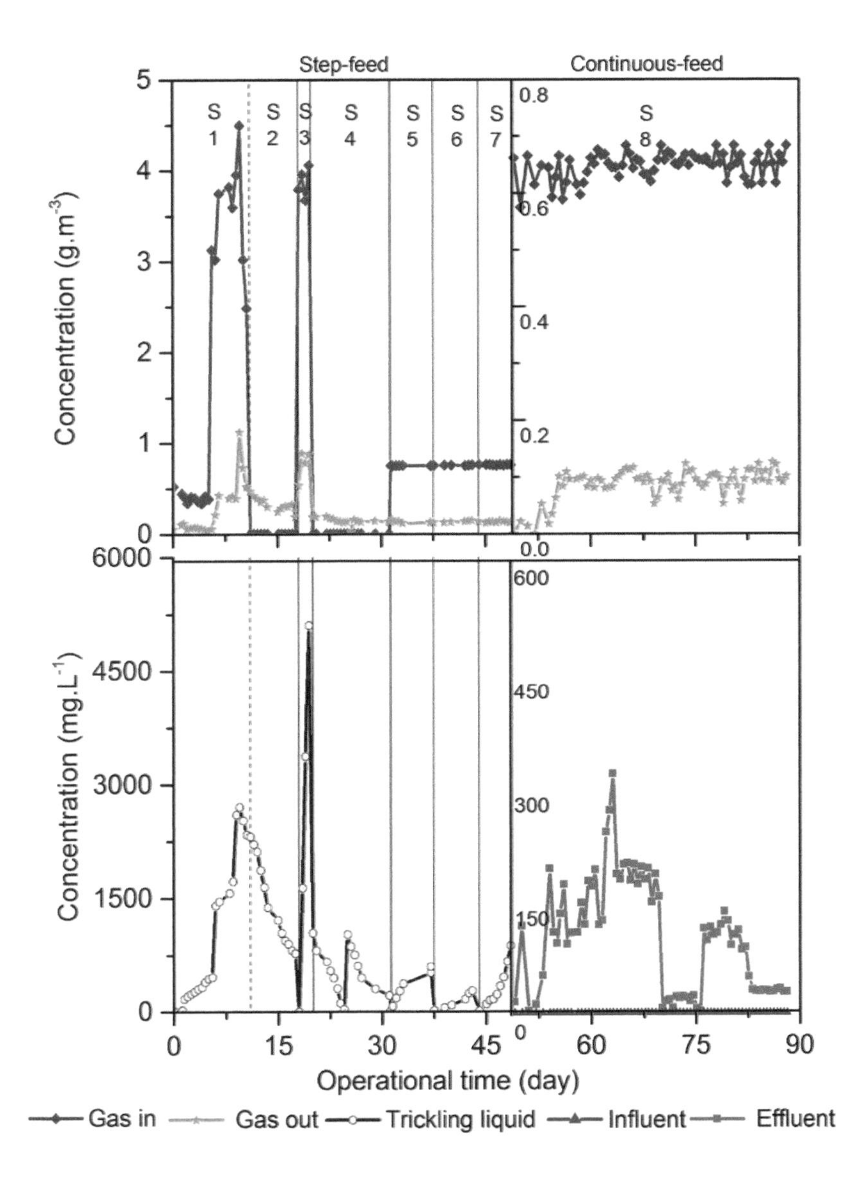

Figure 5.4: *Methanol concentration in the BTF during different phases of operation, (a) inlet and outlet concentration of gas-phase methanol, (b) liquid-phase methanol concentration during step-feed of* SeO_4^{2-} *(in black); influent (in blue) and effluent (in red) methanol concentration during continuous-feed of* SeO_4^{2-} *without trickling liquid recirculation indicating accumulation of up to 320 mg/L of methanol.*

Table 5.2: *Overview of the* $SeOx^{2-}$ *reduction studies reported in the literature.*

$SeOx^{2-}$ concentration		Reactor configuration	Electron donor	$SeOx^{2-}$ removal efficiency	References
Selenate: Step feed: ~ 50 $mg.L^{-1}$ Continuous feed: 7-9 $mg.L^{-1}$	Anaerobic reduction	Biotrickling filter (BTF)	Methanol: 5-55 $g.m^{-3}.h^{-1}$ or 1000 $mg.L^{-1}$	Step feed: > 90% at 7.5 $mg.L^{-1}.d^{-1}$ Continuous feed: ~68% at 4.5 $mg.L^{-1}.d^{-1}$	This study
Selenite: 12.7 $mg.L^{-1}$	Aerobic reduction	Continuous-flow activated sludge reactor	Glucose: 1000 $mg.L^{-1}$	86 %	Jain et al. (2016)
Selenate: 7.2 $mg.L^{-1}$	Anaerobic reduction with co-reduction of nitrate	Upflow anaerobic sludge blanket (UASB) reactor	Lactate: 125 $mg.L^{-1}$	Thermophilic: 94.4% Mesophilic: 82%	Dessì et al. (2016)
Selenate: 18.7 $mg.L^{-1}$	Anaerobic reduction with nitrate and sulfate reduction	Drip flow reactors	Lactate: 1800 $mg.L^{-1}$	Only SeO_4^{2-} : 30% $NO_3^-+SeO_4^{2-}$: 37% $SO_4^{2-}+SeO_4^{2-}$: 77% $NO_3^-+SO_4^{2-}+SeO_4^{2-}$:76%	Tan et al. (2018)
Selenate: 9.15 $mg.L^{-1}$	Anaerobic reduction by denitrifying anaerobic methane oxidizing biofilm	Membrane biofilm reactor	Methane: 18.8 $L.m^{-2}.d$	62.5%	Luo et al. (2018)
Selenate: 1 $mg.L^{-1}$	Anaerobic selenate and nitrate reduction	Membrane biofilm reactor	Methane: in excess	60-100%	Lai et al. (2016)

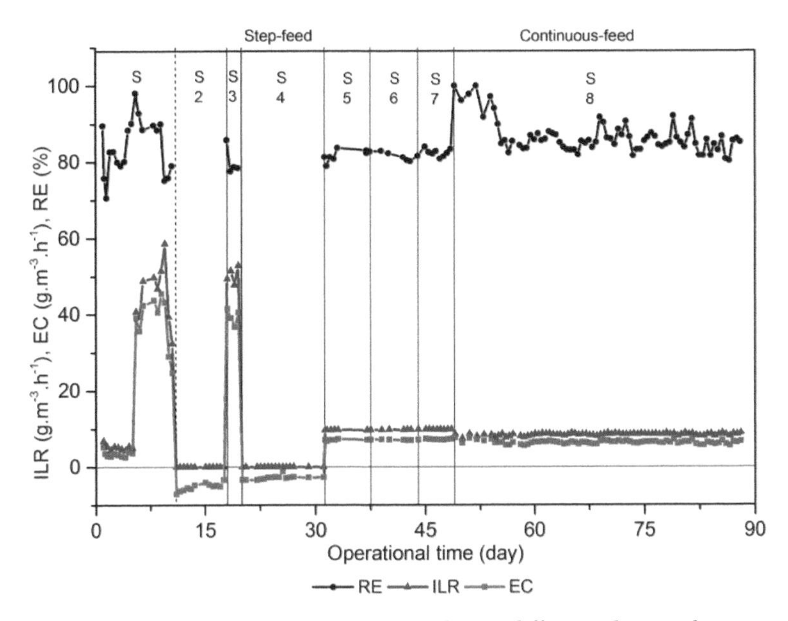

Figure 5.5: *Methanol removal in the BTF during different phases of operation with gas-phase methanol removal efficiency (RE) in black, elimination capacity (EC) and inlet loading rate (ILR) for gas-phase methanol in red and blue, respectively.*

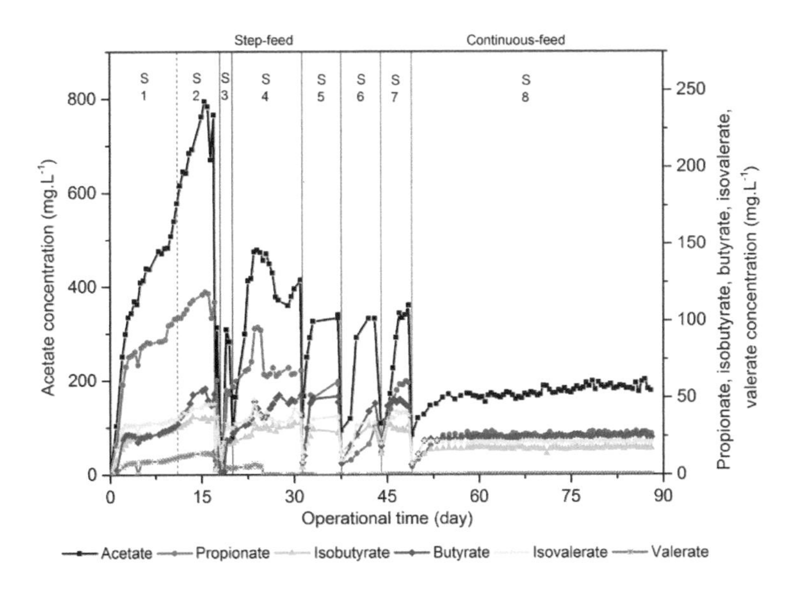

Figure 5.6*: Concentrations of acetate, propionate, isobutyrate, butyrate, isovalerate and valerate in the trickling liquid during various phases of BTF operation. Vertical lines indicate the change of phase: solid lines and dotted lines indicate the continuation of the subsequent phase, respectively, with and without changing the trickling liquid.*

5.4. Discussion

5.4.1. Selenate reduction in the methanol fed BTF

This study shows that SeO_4^{2-} can be removed from wastewaters in a BTF via its reduction using gas-phase methanol as the electron donor. The increase in the average SeRR from 2.8 $mg.L^{-1}.d^{-1}$ in phases S1 and S2 to 9.9 $mg.L^{-1}.d^{-1}$ in phase S7 indicates an enrichment of SeO_4^{2-} reducing microorganisms in the biomass with the progressive BTF operation. The average SeO_4^{2-} reduction efficiency was > 90% and ~ 70% at a rate of, respectively, 7.5 and 4.5 $mg.L^{-1}.d^{-1}$ during the step and continuous feed operations. For SeO_4^{2-} removal, the performance of the BTF was better during the step-feed than the continuous-feed mode. Thus, for effluents containing high SeO_x^{2-} concentrations, such as those from selenium, gold and lead mining industries (that contain up 620, 33 and 7 $mg.L^{-1}$ SeO_x^{2-}, respectively) or oil refining industries (up to 5 $mg.L^{-1}$) (Pettine et al., 2015; Tan et al., 2016), the step-feed mode of operation is beneficial in terms of Se bioreduction, although this might incur higher operational costs.

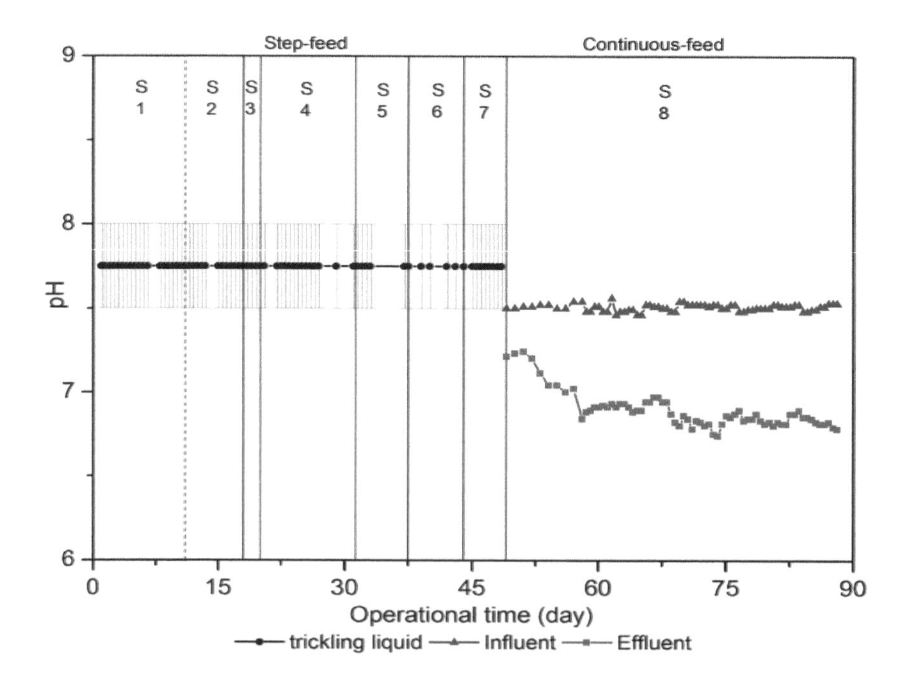

Figure 5.7: *pH profile of the trickling liquid during different phases of BTF operation.*

In step-feed operation, the effluent SeO_4^{2-} concentration could be regulated by controlling the duration of the BTF operation. For instance, in phase S4, the effluent SeO_4^{2-} concentration was < 0.2 $mg.L^{-1}$ with a RE > 98%. However, achieving the stringent effluent discharge limits for Se, e.g. 0.001–0.1 $mg.L^{-1}$ in countries like Korea, Canada and Japan (Tan et al., 2016), is rather difficult and operational parameters like increased RT, lower SeO_4^{2-} loading rate or a post-treatment process should be considered to improve the effluent quality. Besides, several studies have already reported the aerobic and anaerobic bioreduction as well as Se recovery at concentrations < 10 $mg.L^{-1}$ (***Table 5.2***).

From the variations of total dissolved Se concentrations over time (***Figure 5.3***), it is apparent that > 80% of the dissolved Se was in the form of SeO_4^{2-}, implying that the reduction of SeO_4^{2-} to Se^0 was the predominant Se conversion pathway. SeO_4^{2-} reduction is carried out by many bacterial and archaeal species commonly present in sludges from wastewater treatment plants, e.g. activated sludge process (Jain et al., 2015) and UASB reactors (Dessì et al., 2016). A number of process studies have demonstrated the aerobic reduction of SeO_x^{2-} to Se^0 using activated sludge from the municipal wastewater treatment plant from where the inoculum for this study was procured (Jain et al., 2016; Mal et al., 2017).

Table 5.3: *EC_{max} values for gas-phase methanol at respective EBRT.*

Gas-phase composition	Reactor type and operation	Methanol ILR ($g.m^{-3}.h^{-1}$)	Methanol EC_{max} ($g.m^{-3}.h^{-1}$)	EBRT (s)	References
Methanol	Anaerobic BTF, thiosulfate rich trickling liquid	25-30	21	138	Eregowda et al. (2018)
Methanol	Anaerobic BTF, selenate rich trickling liquid	7-56	46	276	This study
Methanol + formaldehyde	Aerobic BF and BTF	Up to 644.1	600	20.7-71.9	Prado et al. (2004)
Methanol + H_2S	Aerobic single stage BTF, low pH	240-260	236	24	Jin et al. (2007)
Methanol + H_2S + α-pinene	Aerobic two stage: BF and BTF	28-1260	894	41.7	Rene et al. (2010)
Methanol	Aerobic BTF	3700	2160	20-65	Ramirez et al. (2009)
Methanol + formaldehyde	Aerobic BTF	Up to 700	552	36	Prado et al. (2006)

Note: ILR-inlet loading rate; EC_{max}-maximum elimination capacity; EBRT-empty bed residence time; BF-biofilter; BTF-biotrickling filter

The concentration of volatile Se in the gas trap was < 0.022% of the Se reduced in phase S8. Although the volatile fraction appears to be insignificant, given the toxicity potential of organo-Se compounds at lower concentrations, Se volatilization still constitutes a health risk in full-scale applications (Lenz and Lens, 2009). However, organo-Se compounds are highly reactive and can be accumulated using an acidic gas trap.

5.4.2. Anaerobic utilization of methanol in the presence of selenate

Methanol is a cheap carbon source and electron donor for the treatment of sulfate (Suhr et al., 2015), nitrate (Fernández-Nava et al., 2010) and zinc (Mayes et al., 2011) rich wastewaters. Anaerobic methylotrophy has been thoroughly investigated in UASB reactors, and the methanol utilization in the absence of nitrate and sulfate was shown to mainly occur by methanogens and acetogens (Fernández-Nava et al., 2010). Since no methanogenic activity was detected during the 89 d of BTF operation, methanol was mainly utilized by acetogens in the BTF investigated.

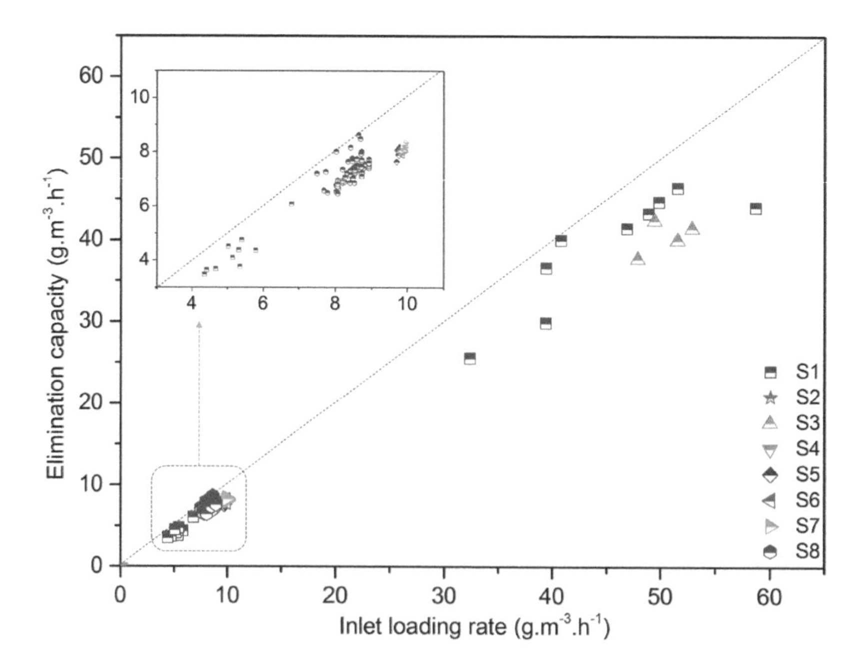

***Figure 5.8**: EC as a function of ILR for gas-phase methanol during different phases of BTF operation (S1-S8).*

The EC of the gas-phase methanol increased linearly with the ILR in phases S1, S5, S6, S7 and S8 (***Figure 5.8***), when the ILR was in the range of 7.7–9.0 $g.m^{-3}.h^{-1}$. An EC_{max} of 46 $g.m^{-3}.h^{-1}$ was achieved in phase S1 with a RE > 80% when the ILR was 50-56 $g.m^{-3}.h^{-1}$. In phases S2 and S4, when the methanol was supplied only in the liquid-phase, an equivalent of EC for the liquid-phase methanol or the methanol removal rate ($mg.L^{-1}.h^{-1}$) was, respectively, 7.6 and 11.7 $mg.L^{-1}.h^{-1}$ for the phases S2 and S4. Several studies have reported the removal of gas-phase methanol in a BF or BTF (***Table 5.3***). Removal of methanol in a BF or BTF can reach an EC_{max} as high as 2160 $g.m^{-3}.h^{-1}$ at an EBRT of 20–65 s (Ramirez et al., 2009). However, these studies were carried out under aerobic conditions and the biomass growth rate for an anaerobic process is lower compared to aerobic processes (Choubert et al., 2009). Thus, lower EC values can be expected for an anaerobically operated BTF.

Eregowda et al. (2018) studied the anaerobic utilization of gas-phase methanol along with (step-fed) thiosulfate reduction in a BTF. Under a similar ILR range of 7.7–9.0 $g.m^{-3}.h^{-1}$, the methanol RE was > 95%, whereas the RE was in the range of 80–90% in this study. The average methanol accumulation was < 75 $mg.L^{-1}$ in the former study, while it was 275–660 $mg.L^{-1}$ (phases S5, S6 and S7) in the present study. Furthermore, during higher methanol loading (40–60 $g.m^{-3}.h^{-1}$) periods in phases S1 and S3, up to 2500 $mg.L^{-1}$ of unutilized methanol accumulated in the liquid-phase. The lower RE and higher methanol accumulation in comparison with Eregowda et al. (2018) suggests that the high SeO_4^{2-} concentrations (up to 60 $mg.L^{-1}$) induced toxicity to the methylotrophic organisms. Jain et al. (2016) also reported the non-recoverable collapse of the continuously operating activated sludge reactor fed with 17.2 $mg.L^{-1}$ of SeO_3^{2-} (Jain et al., 2016).

5.4.3. Practical implications

Bioreduction of SeO_x^{2-} in a BTF along with anaerobic oxidation of methanol could be extended for industrial effluents from the mining, coal firing and pulp and paper industries. For example, the anaerobic treatment of the liquid effluent of a flue gas desulfurization unit containing ~ 10 $mg.L^{-1}$ of Se (Santos et al., 2015) with methanol rich non-condensable gases (NCG) produced during chemical pulping in the pulp and paper industry would reduce the number of process steps and treatment costs compared to individual bioprocesses to treat the pollutants separately. However, further studies on the reduction of SeO_x^{2-} in the presence of other oxyanions such as sulfate, nitrate, phosphate, which are common components of wastewaters from flue gas desulfurization along with oxidation of volatile compounds commonly found in NCG (such as, methyl mercaptan, dimethyl sulfide, dimethyl disulfide, acetone, ethanol and α-pinene) are required to further understand the process stability and end-products formed.

The effluent Se concentration was higher than the industrial discharge limits (< 1 $mg.L^{-1}$). This could be overcome by including a post-treatment step such as the use of another bioreactor configuration (UASB or fed batch system) or a physico-chemical process like adsorption (Howarth et al., 2015). Further studies to assess the microbial community using DNA/RNA sequencing to identify the dominant species, resilience of the biomass towards Se toxicity and Se accumulation in the filter bed can help to further optimize the methanol driven SeO_x^{2-} removal process.

5.5. Conclusions

The use of continuous gas-phase methanol supply for the reduction of SeO_4^{2-} was successfully demonstrated in a BTF under step (up to 60 $mg.L^{-1}$) and continuous (~ 7 $mg.L^{-1}$) feeding modes. When comparing the SeRR, the step-feed of SeO_4^{2-} was better than the continuous-feed of SeO_4^{2-}. This study, for the first time, provides more insights on the bioreduction of Se (up to 60 $mg.L^{-1}$) laden wastewater using gas-phase methanol in a BTF. For the gas-phase methanol, an EC_{max} of 46 $g.m^{-3}.h^{-1}$ and a RE > 80% was achieved and up to 5000 $mg.L^{-1}$ of methanol accumulated in the liquid-phase. Acetogenesis was the predominant methanol utilization pathway which resulted in the production of up to 800 $mg.L^{-1}$ of acetate and 117 $mg.L^{-1}$ of propionate.

Acknowledgements

We thank Mr. Frank Wiegman (UNESCO-IHE, the Netherlands) for the analytical support during reactor operation and Harnaschpolder WWTP (Delft, the Netherlands) for providing the activated sludge. This work was supported by the Marie Skłodowska-Curie European Joint Doctorate (EJD) in Advanced Biological Waste-to-Energy Technologies (ABWET) funded from Horizon 2020 under the grant agreement no. 643071.

References

Bhattarai, S., Cassarini, C., Rene, E.R., Zhang, Y., Esposito, G., Lens, P.N.L., 2018. Enrichment of sulfate reducing anaerobic methane oxidizing community dominated by ANME-1 from Ginsburg Mud Volcano (Gulf of Cadiz) sediment in a biotrickling filter. Bioresour. Technol. 259, 433–441.

Cassarini, C., Rene, E.R., Bhattarai, S., Esposito, G., Lens, P.N.L., 2017. Anaerobic oxidation of methane coupled to thiosulfate reduction in a biotrickling filter. Bioresour. Technol. 240, 214–222.

Choubert, J.M., Marquot, A., Stricker, A.E., Racault, Y., Gillot, S., Héduit, A., 2009. Anoxic and aerobic values for the yield coefficient of the heterotrophic biomass: Determination at full-scale plants and consequences on simulations. Water SA 35, 103–110.

Deshusses, M.A., Cox, H.H.J., 2003. Biotrickling Filters for Air Pollution Control. Encyclopedia of Environmental Microbiology, California, USA.

Dessì, P., Jain, R., Singh, S., Seder-Colomina, M., van Hullebusch, E.D., Rene, E.R., Ahammad, S.Z., Carucci, A., Lens, P.N.L., 2016. Effect of temperature on selenium removal from wastewater by UASB reactors. Water Res. 94, 146–154.

DSM 63, Leibniz Institut DSMZ-Deutsche Sammlung von Mikroorganismen und Zellkulturen GmbH ; Curators of the DSMZ.

Eregowda, T., Matanhike, L., Rene, E.R., Lens, P.N.L., 2018. Performance of a biotrickling filter for anaerobic utilization of gas-phase methanol coupled to thiosulphate reduction and resource recovery through volatile fatty acids production. Bioresour. Technol. 263, 591–600.

Espinosa-Ortiz, E.J., Rene, E.R., Guyot, F., van Hullebusch, E.D., Lens, P.N.L., 2017. Biomineralization of tellurium and selenium-tellurium nanoparticles by the white-rot fungus *Phanerochaete chrysosporium*. Int. Biodeterior. Biodegrad. 124, 258–266.

Fernández-Nava, Y., Marañón, E., Soons, J., Castrillón, L., 2010. Denitrification of high nitrate concentration wastewater using alternative carbon sources. J. Hazard. Mater. 173, 682–688.

Gabos, M.B., Alleoni, L.R.F., Abreu, C.A., 2014. Background levels of selenium in some selected Brazilian tropical soils. J. Geochemical Explor. 145, 35–39.

Gabriel, D., Deshusses, M.A., 2003. Performance of a full-scale biotrickling filter treating H2S at a gas contact time of 1.6 to 2.2 seconds. Environ. Prog. 22, 111–118.

Hageman, S.P.W., van der Weijden, R.D., Stams, A.J.M., van Cappellen, P., Buisman, C.J.N., 2017. Microbial selenium sulfide reduction for selenium recovery from wastewater. J. Hazard. Mater. 329, 110–119.

Howarth, A.J., Katz, M.J., Wang, T.C., Platero-Prats, A.E., Chapman, K.W., Hupp, J.T., Farha, O.K., 2015. High efficiency adsorption and removal of selenate and selenite from water using metal-organic frameworks. J. Am. Chem. Soc. 137, 7488–7494.

Jain, R., Seder-Colomina, M., Jordan, N., Dessi, P., Cosmidis, J., van Hullebusch, E.D., Weiss, S., Farges, F., Lens, P.N.L., 2015. Entrapped elemental selenium nanoparticles affect physicochemical properties of selenium fed activated sludge. J. Hazard. Mater. 295, 193–200.

Jain, R., Seder-Colomina, M., Jordan, N., Dessi, P., Cosmidis, J., van Hullebusch, E.D., Weiss, S., Farges, F., Lens, P.N.L., 2015. Entrapped elemental selenium nanoparticles affect physicochemical properties of selenium fed activated sludge. J. Hazard. Mater. 295, 193–200.

Jain, R., Matassa, S., Singh, S., van Hullebusch, E.D., Esposito, G., Lens, P.N.L., 2016. Reduction of selenite to elemental selenium nanoparticles by activated sludge. Environ. Sci. Pollut. Res. 23, 1193–1202.

Jin, Y., Veiga, M.C., Kennes, C., 2007. Co-treatment of hydrogen sulfide and methanol in a single-stage biotrickling filter under acidic conditions. Chemosphere 68, 1186–1193.

Khoei, N.S., Lampis, S., Zonaro, E., Yrjälä, K., Bernardi, P., Vallini, G., 2017. Insights into selenite reduction and biogenesis of elemental selenium nanoparticles by two environmental isolates of *Burkholderia fungorum*. N. Biotechnol. 1–11.

Lai, C.Y., Wen, L.L., Shi, L.D., Zhao, K.K., Wang, Y.Q., Yang, X., Rittmann, B.E., Zhou, C., Tang, Y., Zheng, P., Zhao, H.P., 2016. Selenate and nitrate bioreductions using methane as the electron donor in a membrane biofilm reactor. Environ. Sci. Technol. 50, 10179–10186.

Lenz, M., Lens, P.N.L., 2009. The essential toxin: The changing perception of selenium in environmental sciences. Sci. Total Environ. 407, 3620–3633.

Luo, J.H., Chen, H., Hu, S., Cai, C., Yuan, Z., Guo, J., 2018. Microbial selenate reduction driven by a denitrifying anaerobic methane oxidation biofilm. Environ. Sci. Technol. 52, 4006–4012.

Mal, J., Nancharaiah, Y.V., Bera, S., Maheshwari, N., van Hullebusch, E.D., Lens, P.N.L., 2017. Biosynthesis of CdSe nanoparticles by anaerobic granular sludge. Environ. Sci. Nano 4, 824–833.

Mal, J., Nancharaiah, Y.V., van Hullebusch, E.D., Lens, P.N.L., 2017. Biological removal of selenate and ammonium by activated sludge in a sequencing batch reactor. Bioresour. Technol. 229, 11–19.

Matanhike, L., 2017. Anaerobic biotrickling filter for the removal of volatile inorganic and organic compounds (Master's thesis). UNESCO-IHE, Delft, the Netherlands.

Mayes, W.M., Davis, J., Silva, V., Jarvis, A.P., 2011. Treatment of zinc-rich acid mine water in low residence time bioreactors incorporating waste shells and methanol dosing. J. Hazard. Mater. 193, 279–287.

Pettine, M., McDonald, T.J., Sohn, M., Anquandah, G.A.K., Zboril, R., Sharma, V.K., 2015. A critical review of selenium analysis in natural water samples. Trends Environ. Anal. Chem. 5, 1–7.

Prado, Ó.J., Veiga, M.C., Kennes, C., 2006. Effect of key parameters on the removal of formaldehyde and methanol in gas-phase biotrickling filters. J. Hazard. Mater. 138, 543–548.

Prado, Ó.J., Veiga, M.C., Kennes, C., 2004. Biofiltration of waste gases containing a mixture of formaldehyde and methanol. Appl. Microbiol. Biotechnol. 65, 235–242.

Ramirez, A.A., Peter Jones, J., Heitz, M., 2009. Control of methanol vapours in a biotrickling filter: Performance analysis and experimental determination of partition coefficient. Bioresour. Technol. 100, 1573–1581.

Rene, E.R., López, M.E., Veiga, M.C., Kennes, C., 2010. Steady- and transient-state operation of a two-stage bioreactor for the treatment of a gaseous mixture of hydrogen sulphide, methanol and α-pinene. J. Chem. Technol. Biotechnol. 85, 336–348.

Santos, S., Ungureanu, G., Boaventura, R., Botelho, C., 2015. Selenium contaminated waters: An overview of analytical methods, treatment options and recent advances in sorption methods. Sci. Total Environ. 521, 246–260.

Siddiqui, N.A., Ziauddin, A., 2011. Emission of non-condensable gases from a pulp and paper mill - a case study. J. Ind. Pollut. Control 27, 93–96.

Soda, S., Kashiwa, M., Kagami, T., Kuroda, M., Yamashita, M., Ike, M., 2011. Laboratory-scale bioreactors for soluble selenium removal from selenium refinery wastewater using anaerobic sludge. Desalination 279, 433–438.

Suhr, M., Klein, G., Kourti, I., Gonzalo, M.R., Santonja, G.G., Roudier, S., Sancho, L.D., 2015. Best available techniques (BAT) - reference document for the production of pulp, paper and board. Eur. Comm. 1-906.

Takada, T., Hirata, M., Kokubu, S., Toorisaka, E., Ozaki, M., Hano, T., 2008. Kinetic study on biological reduction of selenium compounds. Process Biochem. 43, 1304–1307.

Tan, L.C., Nancharaiah, Y.V, van Hullebusch, E.D., Lens, P.N.L., 2016. Selenium: environmental significance, pollution, and biological treatment technologies. Biotechnol. Adv. 34, 886-907.

Tan, L.C., Espinosa-Ortiz, E.J., Nancharaiah, Y. V, van Hullebusch, E.D., Gerlach, R., Lens, P.N.L., 2018. Selenate removal in biofilm systems: effect of nitrate and sulfate on selenium removal efficiency, biofilm structure and microbial community. J. Chem. Technol. Biotechnol. 93, 2380–2389

Tang, C., Huang, Y.H., Zeng, H., Zhang, Z., 2014. Reductive removal of selenate by zero-valent iron: The roles of aqueous Fe^{2+} and corrosion products, and selenate removal mechanisms. Water Res. 67, 166–174.

Wadgaonkar, S.L., Mal, J., Nancharaiah, Y.V., Maheshwari, N.O., Esposito, G., Lens, P.N.L., 2018. Formation of Se(0), Te(0), and Se(0)–Te(0) nanostructures during simultaneous bioreduction of selenite and tellurite in a UASB reactor. Appl. Microbiol. Biotechnol. 102, 2899–2911.

Winkel, L.H.E., Johnson, C.A., Lenz, M., Grundl, T., Leupin, O.X., Amini, M., Charlet, L., 2012. Environmental selenium research: From microscopic processes to global understanding. Environ. Sci. Technol. 46, 571–579.

CHAPTER 6

Selenate bioreduction using methane as the electron donor in a biotrickling filter

Abstract

This study investigated anaerobic bioreduction of selenate to elemental selenium by marine lake sediment from Lake Grevelingen in the presence of methane as a sole electron donor in both batch and continuous studies. Complete bioreduction of 14.3 $mg.L^{-1}$ selenate was achieved in batch studies, while up to 140 $mg.L^{-1}$ selenate was reduced in the biotrickling filter with a continuous supply of 100% methane gas (in excess) at a reduction rate of 4.12 $mg.L^{-1}$ d^{-1}. Red coloured deposits, characteristic of Se(0) particles were observed inside the polyurethane foam packing in the biotrickling filter, which were further confirmed using graphite furnace atomic absorption spectrophotometer. The biotrickling filter was operational for 348 days and > 97% selenate reduction was observed for the entire time required for complete bioreduction of selenate decreased gradually with each step feed of selenate.

Keywords: Biotrickling filter, selenate, bioreduction, methane, marine lake Grevelingen sediment, selenium wastewater

6.1. Introduction

A range of anthropogenic activities like mining, metallurgy and refining, production of photoelectric devices and semiconductors, power generation and agriculture along with natural weathering and soil leaching contribute towards the enrichment of inorganic selenium (Se) in water streams (Chapman et al., 2009; Tan et al., 2016). Selenium is an essential micronutrient for humans and animals due to its importance in metabolic pathways and immune functions (Zimmerman et al., 2015) and its role in the antioxidant glutathione peroxidase, which protects the cell membrane from damage caused by peroxidation of lipids (Tinggi, 2003). Although Se is an essential trace mineral, an intake at too high concentrations is detrimental leading to Se toxicity (Macfarquhar et al., 2011; Tan et al., 2016). Furthermore, its high rate of bioaccumulation at many trophic levels, even at lower concentrations, makes it a serious human health and environmental concern (Bleiman and Mishael, 2010; C. Y. Lai et al., 2016; Tan et al., 2016). Chief sources of Se contamination are coal mining and processing, uranium mining, petroleum extraction and refineries and power production (Lemly, 2004; Muscatello et al., 2008). Selenium contamination due to wastewater discharge from mining industry in marine sediment is a recent environmental concern (Ellwood et al., 2016).

Bioreduction of toxic and soluble Se oxyanions to non-toxic and water insoluble Se(0) forms is a promising approach toward tackling the problem of increasing Se contamination in water bodies. However, present studies (Jain et al., 2016, 2015) employ expensive sources of electron donors for selenium bioreduction. It is necessary to explore cheaper and renewable sources of electron donor for selenium bioreduction on larger scale. Methane is one such potential electron donor.

Methane not only occurs in the form of natural gas, but can also be synthesised renewably (Amon et al., 2007). The International Energy Statistics comprehends that proven reserves stood at 194 trillion cubic metres by the end of 2016 (U.S. E.I.A., 2010). Apart from being a potential energy source, methane is also a potent greenhouse gas (GHG) with a greenhouse warming potential 21 times than that of carbon dioxide (Haynes and Gonzalez, 2014). Anthropogenic sources of methane contribute nearly 20% to the world's GHG warming potential each year when converted to CO_2 (EPA, 2011). Venting and inefficient flaring of natural gas produced as a by-product of petroleum extraction also accounts for 140 trillion cubic meters of GHG released into the atmosphere worldwide in 2011 (Haynes and Gonzalez, 2014). Thus, methane is a cheap and abundantly available carbon source and electron donor for reduction of common sulfur and selenium oxyanions (Bhattarai et al., 2018a; Cassarini et al., 2017; C.-Y. Lai et al., 2016).

The sediment biomass used in this study was collected from the marine lake Grevelingen (Scharendijke basin, the Netherlands), from a sulfate-methane transition zone (SMTZ), a region in the sea sediment where the methane rising from below and the sulfate sinking from above form a region suitable for anaerobic methanotrophy (Mcglynn et al., 2015). Recently Bhattarai et al. (2017) studied and confirmed deep sea sediments from SMTZ for possibilities of the reduction of thiosulfate ($S_2O_3^{2-}$), sulfate (SO_4^{2-}) and sulfur (S^0) using ethanol, lactate, acetate and methane as electron donor. Se has a chemical behaviour similar to that of sulfur. Therefore, a selenate reduction similar to sulfate reduction coupled with methane oxidation using the Grevelingen biomass (Bhattarai et al., 2018a; Cassarini et al., 2017) was expected in this study.

This research was framed to study anaerobic degradation of selenate in batch bottles using methane and acetate individually as electron donors. Furthermore, a continuous system for bioreduction of selenate with methane as the electron donor and carbon source in a BTF was analysed.

6.2. Materials and methods

6.2.1. Biomass collection and medium composition

The sediment sample was collected from the marine lake Grevelingen (MLG), a former Rhine-Meuse estuary on the border of the Dutch provinces of South Holland and Zeeland that has become a lake due to the delta works (Bhattarai et al., 2018a). The sampling site had a salinity of 31.7% and sulfate concentration of 25 mM at the surface of the sediment and sulfate concentration reduced to ~5 mM at a depth of 35 cm (Egger et al., 2016). The sediment layer of 10-20 cm depth was homogenised and used as the inoculum. Total Se analysis using a graphite furnace-atomic absorption spectroscopy (section 6.2.4) showed the total Se concentration in the sediment to be non-detectable (< 0.2 μg/L).

The artificial seawater medium used for the growth of MLG sediment, consisted of NaCl (26 g.L^{-1}), $MgCl_2 \cdot 6H_2O$ (5 g.L^{-1}), $CaCl_2 \cdot 2H_2O$ (1.4 g.L^{-1}), NH_4Cl (0.3 g.L^{-1}), KH_2PO_4 (0.1 g.L^{-1}) and KCl (0.5 g.L^{-1}) (Zhang et al., 2010), with 1 mL.L^{-1} resazurin solution as redox indicator (0.5 g.L^{-1}). The solution with resazurin is colourless at a redox below -110 mV and turns pink at a redox above -51 mV. The following sterile stock solutions described by Widdel and Bak

(Widdel and Bak, 1992) were added to the artificial seawater media: trace element solution (1 mL.L^{-1}), 1 M $NaHCO_3$ (30 mL.L^{-1}), vitamin mixture (1 mL.L^{-1}), thiamine solution (1 mL.L^{-1}), vitamin B12 solution (1 mL.L^{-1}). The trace element solution comprised of ethylene di amine tetra acetic acid (EDTA) disodium salt (5.2 g.L^{-1}), H_3BO_3 (10 mg.L^{-1}), $MnCl_2 \cdot 4H_2O$ (5 mg.L^{-1}), $FeSO_4 \cdot 7H_2O$ (2.1 g.L^{-1}), $CoCl_2 \cdot 6H_2O$ (190 mg.L^{-1}), $NiCl_2 \cdot 6H_2O$ (24 mg.L^{-1}), $CuCl_2 \cdot 2H_2O$ (10 mg.L^{-1}), $ZnSO_4 \cdot 7H_2O$ (144 mg.L^{-1}), $Na_2MoO_4 \cdot 2H_2O$ (36 mg.L^{-1}) and the pH of the medium was adjusted to 7.0 (±0.1). The mineral medium was flushed and pressurized with N_2 to remove oxygen (Fermoso et al., 2008).

6.2.2. Batch experiments

Selenate reduction by marine sediment inoculum using different electron donors

Batch studies for selenate reduction were carried out in triplicates by transferring 200 mL of sterile artificial sea water medium in 250 mL airtight serum bottles with thick butyl rubber septa and sealed with aluminium crimps. The serum bottles were made anoxic by purging nitrogen and seeded with 5 mL of the sediment as the source of biomass in an anaerobic chamber.

Different controls containing no selenate, no sediment, heat-killed sediment and no electron donor were prepared in triplicates. An initial concentration of 14.3 mg.L^{-1} of selenate was added to each bottle except for the control without selenate. For batches containing acetate as the electron donor, 250 mg.L^{-1} of acetate was added with nitrogen in the head space at 2 bar; whereas for those containing methane as the electron donor, 100% methane was filled in the headspace at a pressure of 2 bars. The control without electron donor was incubated with 100% nitrogen in the head space at 2 bars. All the batch bottles were incubated in an orbital shaker (Cole-Parmer, Germany) at 200 rpm in a temperature regulated room temperature (20±1) in the dark for a period of 77 days. The liquid samples were analysed for selenate, total selenium, volatile fatty acids (VFA) and gas samples for methane and carbon dioxide every 2 weeks.

Selenate reduction in the presence of ^{13}C-labelled methane ($^{13}CH_4$)

In order to confirm the occurrence of anaerobic methane oxidation coupled with selenate reduction, the methane oxidation to CO_2 was monitored in the presence of ^{13}C-labelled methane ($^{13}CH_4$) as electron donor. The production of ^{13}C-labelled carbon dioxide ($^{13}CO_2$) was investigated along with selenate removal in batch tests, in the presence of methane as sole electron donor comprising of 5% $^{13}CH_4$ and 95% $^{12}CH_4$. The tests were carried out at atmospheric pressure along with different controls in triplicated as suggested in the previous batch test.

6.2.3. Biotrickling filter

A laboratory scale BTF outlined in ***Figure 6.1*** was constructed to investigate anaerobic selenate bioreduction using methane as an electron donor and consisted of a filter bed encompassed in an airtight glass cylindrical column of 12 cm inner diameter and 50 cm packing height, with a volume of 4.6 L, provided with adequate sampling ports for biomass, gas and liquid sampling. The column was packed with polyurethane foam (98% porosity and a density of 28 $kg.m^{-3}$) cut in cubes of 1-1.5 cm^3, mixed with an equal ratio of plastic pall rings (specific surface area of 188 $m^2.m^{-3}$ and bulk density of 141 $kg.m^{-3}$). The bed was supported by two perforated support plates with 2 mm openings.

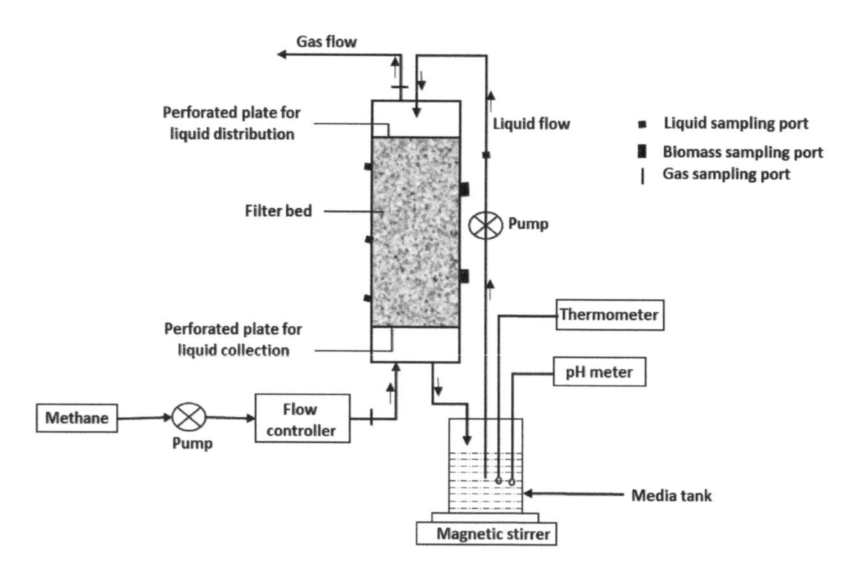

Figure 6.1. *Schematic representation of the BTF system for anaerobic reduction of selenate as step feed using methane as electron donor and carbon source.*

The fluids were circulated in a counter flow with 100% methane in gas-phase passed through the column from bottom to top by means of a mass flow controller (SLA 5800 MFC, Brooks instruments) as a continuous feed at a rate of 180 $mL.h^{-1}$. The estimated empty bed residence time for methane was 25.5 h. The artificial seawater medium trickled from the top at a rate of 7.2 $L.h^{-1}$, with a residence time of 1.56 h. The liquid drain was collected into an airtight 3 L glass tank equipped with a mechanical stirrer and a pH probe (pH/mV transmitter - Delta OHM

DO9785T) and recirculated to the filter bed through a peristaltic pump (Masterflex L/S Easy-Load II, model 77201-60). The operational parameters are as summarised in ***Table 6.1***.

The biotrickling filter was seeded with 500 mL sediment (50% fresh and 50% enriched biomass from the batch studies). 14.3 $mg.L^{-1}$ of selenate was added to artificial sea water media in phase one, followed by a 1154 $mg.L^{-1}$ of sulfate in phase II to enhance the sulfate reducing activity in the biomass, (Hockin and Gadd, 2003; Y V Nancharaiah and Lens, 2015; Zehr and Oremland, 1987). The concentration of the selenate step-fed to the reactor was gradually increased in each phase as summarised in ***Table 6.2***.

Table 6.1. *Operational parameters of the biotrickling filter.*

Operational parameters	Conditions
Temperature (°C)	20
pH	7-8
Electron acceptor ($mg.L^{-1}$)	Selenate
Carbon source and electron donor	Methane
Growth medium	Artificial seawater
Nutrient loading rate ($mL.min^{-1}$)	250
Inlet gas flow rate ($mL.h^{-1}$)	180
Volume of BTF bed (L)	4.6
Packing material	Polyurethane foam cubes + plastic pall rings
BTF bed height (cm)	90
Media change	At the beginning of each phase

After 348 days of reactor operation, the Se(0) accumulation in the polyurethane foam cubes was analysed. The foam cubes were washed in Milli-Q water to remove unbound Se and dried in a hot air oven at 70 °C overnight and the weight was recorded. The dry foam cube was cut into tiny pieces of 0.3 – 0.5 mm width, washed in 10 mL 20% HNO_3 (v/v). Washing was repeated serially 10 times. The 10th wash water was analysed for total Se and showed negative for Se, thus indicating that Se was completely extracted from the polyurethane foam cube pieces by serial washing in acidified water. The wash waters from all the 10 washes were mixed together and sonicated at a density of 0.52 W mL^{-1} for 3 minutes to disrupt intact biomass. The liquid sample was then analysed for total Se.

Table 6.2. *Phases of operation based on the addition of electron acceptor and selenate removal rates.*

Operation phase	Days	Selenate ($mg.L^{-1}$)	Sulfate ($mg.L^{-1}$)	Methane ($g\ m^{-3}$)	Selenate removal rate ($mg.L^{-1}\ d^{-1}$)
P1	0-24	13.2	0	715.75	1.12
P2	25-56	0	1154.3	715.75	NA
P3	57-97	132	0	715.75	4.12
P4	98-249	13.2	0	715.75	0.79
P5	250-348	39.6	0	715.75	1.45

6.2.4. Analytical methods

The parameters under consideration for this study were selenate, sulfate, CO_2, methane, VFA and total Se. Selenate and sulfate were analysed using ion chromatography (ICS-1000, IC, Dionex with AS-DV sampler) as described by (Cassarini et al., 2017). Methane and CO_2 in the batch bottles and BTF were measured by gas chromatograph (Scion Instruments 456-GC) fitted with a thermal conductivity detector using He as carrier gas at 1.035 bars as described by Bhattarai et al. (2017). The VFA were measured by gas chromatograph (Varian 430-GC) with helium as a carrier gas at 1.72 bars. The total Se was analysed in a GF-AAS (Solaar AA Spectrometer GF95) as described by (Mal et al., 2016).

Selenate reduction rate and methane utilization were calculated as follows:

The volumetric selenate reduction rate (SeRR) for the BTF at different phases was calculated as follows:

$$\text{SeRR} = \frac{[SeO_4^{2-}]_t - [SeO_4^{2-}]_{(t+\Delta t)}}{\Delta t} \quad \text{(Eq. 6.1)}$$

Percentage methane utilization in the BTF was calculated as follows:

$$\text{Methane utilization (\%)} = \frac{Gas_{in} - Gas_{out}}{Gas_{in}} \times 100 \quad \text{(Eq. 6.2)}$$

Where $[SeO_4^{2-}]$ = concentration of selenate ($mg.L^{-1}$); t, Δt = time (d); Gas_{in} = Area under the curve for the inlet gas of the BTF in $mV.min^{-1}$; Gas_{out} = Area under the curve for the gas outlet of the BTF in $mV.min^{-1}$.

6.3. Results

6.3.1. Selenate reduction in batch studies

Anaerobic selenate removal (Eq. 6.1) by MLG sediment was monitored as a function of time with methane as electron donor along with the controls. Complete bioreduction of 14.3 mg.L^{-1} of selenate was observed in about 77 days in the bottles with live biomass, while the controls with no methane and killed biomass, both showed a selenate removal of about 37% (***Figure 6.2***). A thin layer of red coloured particles was observed only in the batch bottles with selenate, methane and live biomass indicating the formation of Se(0) particles (***Figure 6.3***). Control batches without the biomass, and with killed biomass did not show red coloured particles.

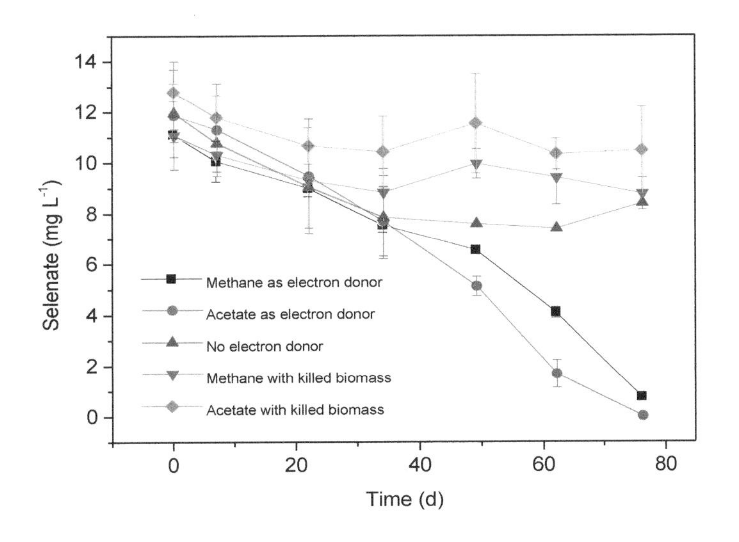

Figure 6.2. *Selenate profiles as a function of time with methane and acetate as electron donors, along with controls (killed biomass and the absence of electron donor) showing complete removal of selenate with both methane and acetate as electron donors using BTF biomass sampled on 76 days.*

A similar trend of selenate reduction was observed in the serum bottles with acetate as the electron donor. Complete reduction of 14.3 mg.L^{-1} of selenate was observed in 80 days in the batches with live biomass, while 23% of selenate removal was observed in the batches with killed biomass and 40.5% in batches with no electron donor without externally supplied electron donor (***Figure 6.4***). For the selenate reduction studies, an initial concentration of 250 mg.L^{-1} of acetate was used and acetate was removed with the course of time, indicating that acetate was indeed used as electron donor for selenate reduction. This is further supported by the fact that acetate was not being consumed in the batches with killed biomass.

Figure 6.3. *Thin layer of red coloured Se(0) particles on the media in the serum batch bottles with 14.3 mg.L^{-1} of selenate, methane in the head space at 2 bars and 5 mL MLG sediment in artificial sea water media.*

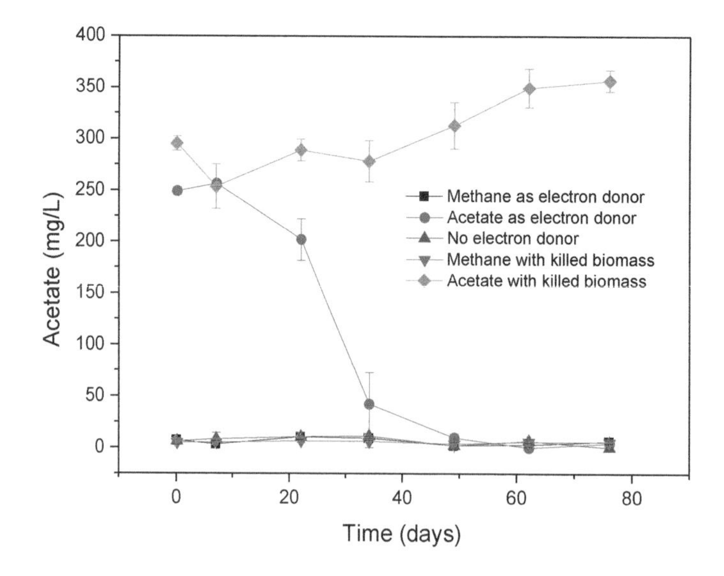

Figure 6.4: *Acetate consumption profile with live and killed biomass showing complete degradation of acetate in 50 days in batches with live biomass and no degradation in batches with autoclaved biomass.*

6.3.1.1. ^{13}C-labelled methane as electron donor

In contrast to the previous experiment, no selenate removal from each batch bottles in the presence of $^{13}CH_4$ as sole source of electron donor was observed (***Figure 6.5a***). On day 7, a steep decrease in concentration of selenate in each bottle was observed. This could be attributed to the absorption of selenate on the sediments or due to sampling/analytical error. No significant change was observed in selenate in medium and Se in sediment observed after 35 days. On day 45 and 55, slight decrease in selenate concentration in the medium was observed in the experimental set-ups with ^{13}C and ^{12}C methane in the headspace but also in control without methane in the headspace (headspace with N_2). This removal of selenate from the medium could be due to oxidation of the organic carbon content in the sediment and not because of oxidation of ^{13}C-labelled methane. Similarly, the changes in the ^{13}C-labelled carbon dioxide (***Figure 6.5b***) was similar to that observed in the test and control bottles, suggesting that the methane was not utilised for the selenate reduction in the batch experiments.

6.3.2. Performance of the biotrickling filter

Selenate and sulfate bioreduction

Selenate bioreduction was evaluated in the biotrickling filter as a function of time by dividing the total study period of 348 days into 5 phases (***Table 6.2***), based on the selenate or sulfate step feed. 100% removal of both selenate and sulfate was achieved in the entire 348 days study. A removal rate ranging from 0.31-4.12 $mg.L^{-1}.d^{-1}$ of selenate and 32.48 $mg.L^{-1}.d^{-1}$ of sulfate were observed (***Figure 6.6***). The average removal rates of selenate for each phase is summarised in ***Table 6.2***. In the phase P4 and P5, the average removal rates for each step feed of selenate increased with each run. Red colour precipitates, characteristic of elemental Se(0) (Lenz et al., 2008), were observed on the sponges at the end of BTF run (***Figure 6.6***).

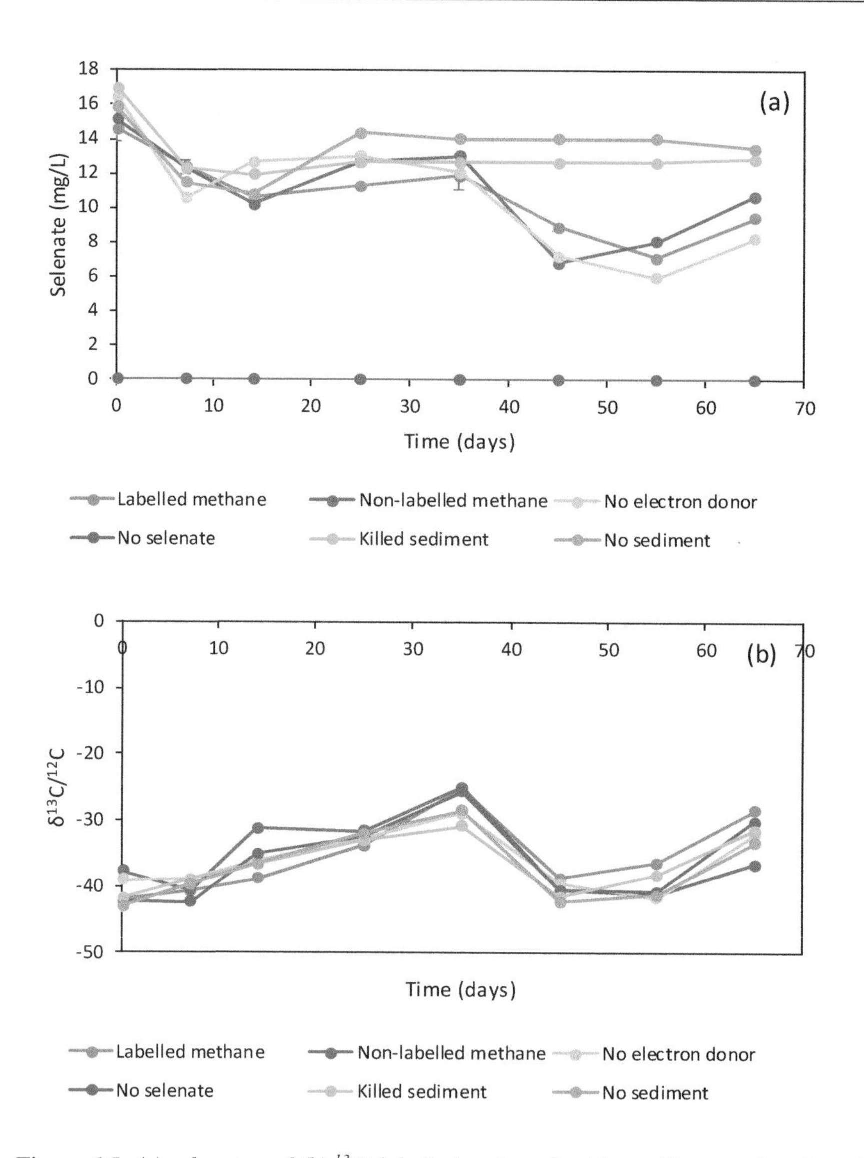

Figure 6.5: *(a) selenate and (b)* ^{13}C*-labelled carbon dioxide profiles as a function of time with* ^{13}C*-labelled methane and* ^{12}C *methane as electron donors, along with controls (killed biomass and the absence of electron donor).*

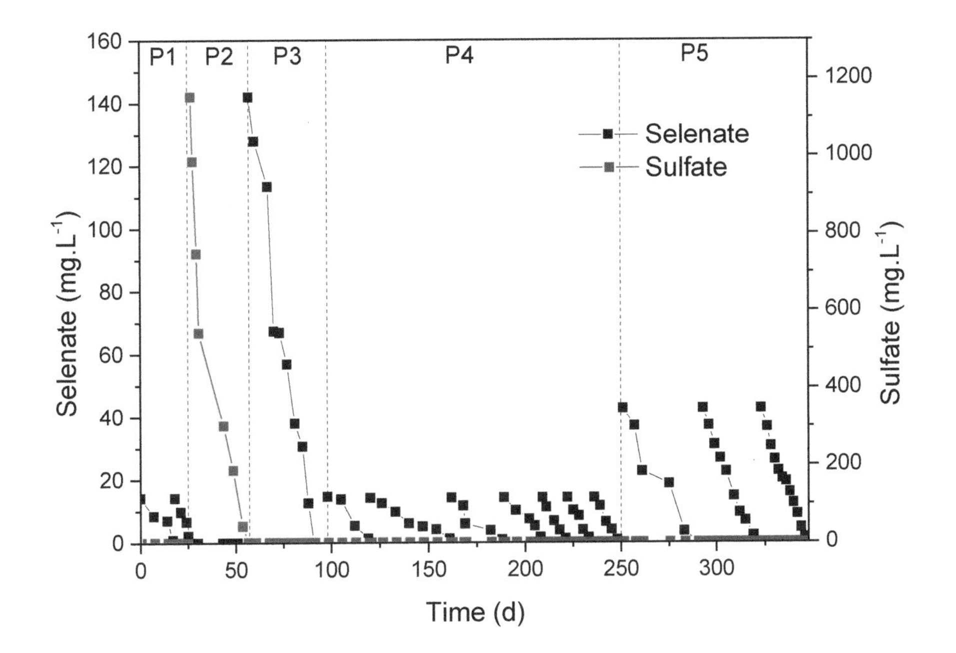

Figure 6.6: *Time profile of selenate and sulfate in the biotrickling filter phased out based on the initial step feed concentration showing complete removal of sulfate and selenate. The vertical lines represent different phases in the BTF operation; P1: days 0-24, P2: days 25-56, P3: days 57-97, P4: days 98-249, P5: days 250-348. Artificial sea water media was changed at the beginning of each phase.*

6.3.1.1. Acetate and propionate production and methane utilization

From the beginning of phase P4 at 98 days of operation, the VFA profiles were monitored to account for the high selenate removal rate of 4.12 $mg.L^{-1}.d^{-1}$ in phase P3. Concentrations up to 60 $mg.L^{-1}$ of propionate were observed which eventually decreased to below detection limit within 175 days of operation. In case of acetate, 580 $mg.L^{-1}$ was observed at the beginning of phase P4 and decreased to 185 $mg.L^{-1}$ after 245 days of operation and eventually to below detection limits after 280 days of operation (***Figure 6.7a, 6.7b***). The batch tests with acetate as electron donor showed that the biomass was capable of using short chain fatty acids as a carbon source (***Figure 6.4***). The inlet and outlet methane concentration profiles were monitored from phase P4 starting from 98 days of operation. The measurements were not made in P1 to P3 because of technical limitations. Average methane utilization (Eq. 6.2) in the reactor was 8.52% with a maximum utilization of 16-6%, at 263 days of operation in phase P5 (***Figure 6.7c***).

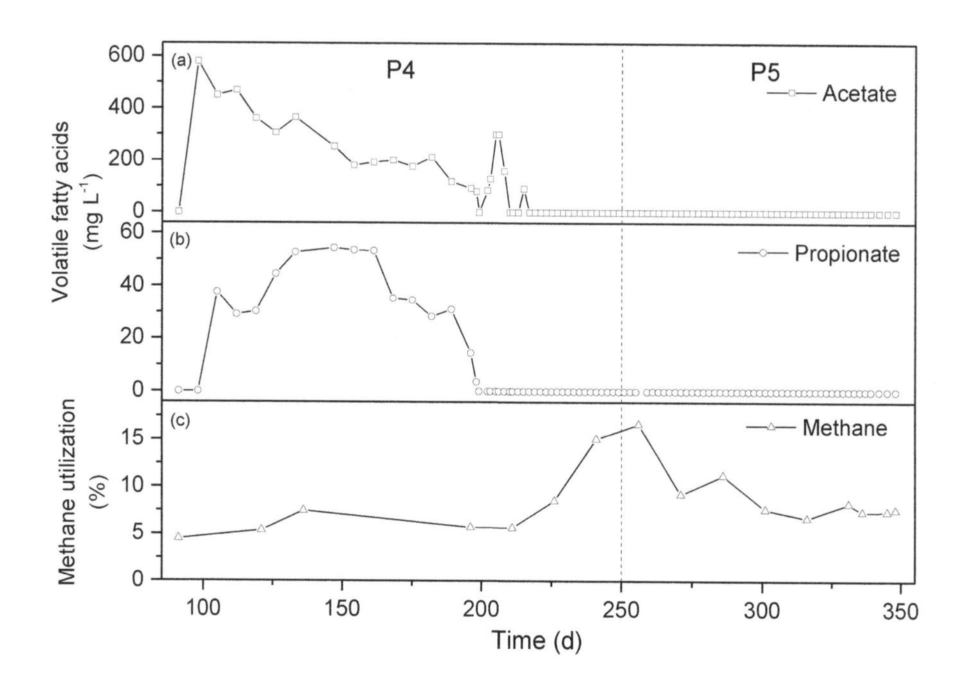

Figure 6.7: *Time profile of (a) acetate and (b) propionate concentrations along with (c) methane utilization (%) in the BTF in phases P4 and P5 The vertical line separates phase P4: days 98-249 and P5: days 250-348.*

6.4. Discussion

6.4.1. Bioreduction of selenate coupled to the anaerobic oxidation of methane

This study concludes for the first time that methane could be used as electron donor for complete bioreduction of selenate in a BTF for a long term operation of 348 days. During the entire 348 days, no operational problems such as biomass over growth, filter clogging was encountered. Selenium oxyanion reduction by chemical and biological processes have been well documented in various studies (Howarth et al., 2015; Nancharaiah and Lens, 2015; Subedi et al., 2017a) that include selenate co-reduction with other electron acceptors like sulfate (Zehr and Oremland, 1987) using different electron donors (Losi et al., 1997; Viamajala et al., 2006). Lai et al. in their work recorded co-bioreduction of selenate and nitrate using methane as an electron acceptor in a membrane biofilm reactor operational for 140 days (Lai et al., 2016). A stable bioreduction of concentrations up to 140 $mg.L^{-1}$ as a shock load and repeated cycles of 14.3 $mg.L^{-1}$ and 42.9 $mg.L^{-1}$ were efficiently reduced without the system crashing and long term operation for 348 days (***Figure 6.6***).

Figure 6.8: *Comparison of polyurethane foam cubes before and after BTF operation: (a) Fresh polyurethane foam cubes before the operation of BTF. (b) Polyurethane foam cubes from the reactor after 348 days of operation showing adsorbed Se(0) in red colour.*

Few studies on reduction of selenium oxyanions explore the possibility of using gas-phase electron donor, especially an inexpensive source like methane (Cassarini et al., 2017) and mostly focus on liquid-phase substrates (Lai et al., 2016; Mal et al., 2017; Nancharaiah and Lens, 2015), while in this work the feasibility of methane as an electron donor was investigated. In recent years, methane has become a very abundantly available resource due to the discovery of new geographical sources and world wide application of anaerobic digestion process (Amon et al., 2007; Leahy et al., 2013). Batch experiments showed that methane was used as the electron donor for complete bioreduction of 14.3 $mg.L^{-1}$ selenate in 77 days (***Figure 6.2***). In contrast, the controls without methane showed only 39.3% reduction. To further confirm the role of the biomass, the controls with no sediment showed 36.6% of selenate bioreduction in the same period of time. This could be attributed to the change in volume in the batches after each sampling. This may be avoided by using sacrificial batches that can be discarded after each sampling.

In order to confirm the process of anaerobic methane oxidation coupled with selenate reduction, ^{13}C-labelled methane was added as sole electron donor in the batch experiments. However, no selenate removal was observed during the process (***Figure 6.5a***). While the previous experiments were performed under 2 bar pressure of $^{12}CH_4$, this study with ^{13}C-labelled methane were performed at atmospheric pressure. Dissolution of methane is significantly affected by the pressure conditions in the batch tests, which could be the possible explanation for no selenate reduction in the batch tests. However, further study with ^{13}C-labelled methane under 2 bar pressure should be carried out to understand the effect of pressure methane utilization for selenate reduction.

Selenate being a component of wastewaters in petroleum refinery waste waters (Leahy et al., 2013; Tan et al., 2016), a process for co-removal, wherein the methane oxidation could complement a process like selenate reduction would be effective. Though the reactor was fed with 100% methane at a rate of 180 mL/h (***Table 6.1***), the bioavailable fraction would be low considering the fact that the filter bed is completely saturated with water and the methane is available for the organisms only after dissolving in the liquid-phase surrounding the biofilm, and the solubility of methane in seawater is 0.055 mg/m^3 (Duan and Mao, 2006; Serra et al., 2006), making the bioavailability of methane a limiting factor.

6.4.2. Acetate and propionate production in the BTF

Acetate is a well-studied substrate for bioreduction of various chalcogen oxyanions (Bhattarai et al., 2018a). The batch results confirm that the marine sediment are capable of selenate reduction (***Figure 6.2***), coupled with acetate utilization (***Figure 6.4***). During the BTF operation, to account for high selenate bioreduction rates in phase P4, the reactor was monitored for VFA production. Acetate and propionate at concentrations of 580 $mg.L^{-1}$ and 185 $mg.L^{-1}$ were observed. This suggests that methane was oxidised to acetate and propionate. Chun-Yu Lai et al. also observed acetate production in their work on selenate reduction using methane as electron donor (C. Y. Lai et al., 2016). Eventually, the concentrations reduced to non-detectable amounts in the course of the long operational period, possibly due to acetate utilization by the biomass for selenate reduction as demonstrated in the batch studies (***Figure 6.4***).

The topic of anaerobic oxidation of methane to acetate and other hydrocarbons is widely discussed as an economically significant process and is becoming attractive for the potential it holds towards alternate fuel technology (Chowdhury and Maranas, 2015; Fei et al., 2014; Kang and Lee, 2016; Nazem-Bokaee et al., 2016). The Thermofischer's process for thermochemical conversion of methane to alcohols and higher hydrocarbons is established and similar conversions (biological) are being explored. Soo et al. in their studies reported the occurrence of reverse methanogenesis and conversion of methane to acetate by expression of enzyme methyl-coenzyme M reductase (Mcr) from ANME-1, cloned in a methanogenic host (Soo et al., 2016). However, the speculation of acetate and propionate production in the reactor needs further confirmation with dedicated studies using labelled methane studies.

6.4.3. Practical implications

In the order of abundance, Se ranks 69th in the earth's crust. Se has many industrial applications like semiconductors and photoelectric cells, metallurgy and glass industries (Macaskie et al., 2010; Tan et al., 2016) and healthcare applications. Increasing the industrial demand of Se may be approached with its recovery from wastewater streams. In this study, an inexpensive and continuous bioprocess for selenium removal and recovery has been explored and must be further pursued on a larger scale. Although, there are a vast number of studies on reduction of Se oxyanions reduction, there is seldom any focus on accumulation and recovery of the reduced form. In this study, Se(0) particles were accumulated in the filter bed and could be recovered. The serial washing of the polyurethane foam cubes in acidified water and analysis for total selenium revealed the accumulation of 3.8 – 8.02 mg Se g^{-1} sponge (dry weight) in the filter bed. Thus, avoiding the carryover of Se further into the environment where oxidation to reduced Se forms is very likely.

Slower growth rate and doubling time of the biomass from Lake Grevelingen led to long term operation of the batches irrespective of the electron donor provided. Previous studies (Espinosa-Ortiz et al., 2015; Liang et al., 2015; Subedi et al., 2017b; Tang et al., 2014) showed removal of similar concentrations of selenium oxyanions within the period of 1-2 weeks. However, the advantage of using MLG sediment is that the biomass is capable of utilizing inexpensive methane as electron donor. Long term and continuous operation of the process using methane as the electron donor might help to increase the reduction rate of the selenium oxyanions and also reduce the operation cost. Methane and selenate being common pollutants from the petrochemical refineries, a process for utilization of methane for bioreduction of selenate, as illustrated in this study, is an interesting option. Furthermore, studies on improving the bioavailability of methane could significantly enhance the methane utilization rate.

6.5. Conclusions

Complete bioreduction of selenate using methane as an electron donor was demonstrated in this study using biomass from a deep sea sediment collected from a sulfur-methane transition zone. Methane is a good candidate for bioreduction of selenate. The rate of bioreduction consistently increased with each cycle of operation and Se(0) accumulated in the filter bed. Microbial community analysis for the enriched biomass would further broaden the understanding of the biochemical pathways, acetate and propionate production and utilization.

References

Amon, T., Amon, B., Kryvoruchko, V., Machmüller, A., Hopfner-Sixt, K., Bodiroza, V., Hrbek, R., Friedel, J., Pötsch, E., Wagentristl, H., Schreiner, M., Zollitsch, W., 2007. Methane production through anaerobic digestion of various energy crops grown in sustainable crop rotations. Bioresour. Technol. 98, 3204–3212.

Balows, A., 1992. The Prokaryotes: a handbook on the biology of bacteria: ecophysiology, isolation, identification, applications. Springer New York, New York, USA.

Bhattarai, S., Cassarini, C., Naangmenyele, Z., Rene, E.R., 2017. Microbial sulfate-reducing activities in anoxic sediment from Marine Lake Grevelingen : screening of electron donors and acceptors. Limnology. doi:10.1007/s10201-017-0516-0

Bleiman, N., Mishael, Y.G., 2010. Selenium removal from drinking water by adsorption to chitosan-clay composites and oxides: Batch and columns tests. J. Hazard. Mater. 183, 590–595.

Cassarini, C., Rene, E.R., Bhattarai, S., Esposito, G., Lens, P.N.L., 2017. Anaerobic oxidation of methane coupled to thiosulfate reduction in a biotrickling filter. Bioresour. Technol. 240, 214–222.

Chapman, P.M., Adams, W.J., Brooks, M.L., Delos, C., Luoma, S.N., Maher, W.A., Olhendorf, H.M., Presser, T.S., Shaw, D.P., 2009. Ecological assessment of selenium in the aquatic environment: Summary of a SETAC Pellston workshop. Pensacola FL (USA).

Chowdhury, A., Maranas, C.D., 2015. Designing overall stoichiometric conversions and intervening metabolic reactions. Sci. Rep. 5, 16009.

Dessì, P., Jain, R., Singh, S., Seder-Colomina, M., van Hullebusch, E.D., Rene, E.R., Ahammad, S.Z., Carucci, A., Lens, P.N.L., 2016. Effect of temperature on selenium removal from wastewater by UASB reactors. Water Res. 94, 146–154.

Dorer, C., Vogt, C., Neu, T.R., Stryhanyuk, H., Richnow, H.H. 2016. Characterization of toluene and ethylbenzene biodegradation under nitrate-, iron (III)-and manganese (IV)-reducing conditions by compound-specific isotope analysis. *Environ. Pollut.,* 211, 271-281.

Duan, Z., Mao, S., 2006. A thermodynamic model for calculating methane solubility, density and gas phase composition of methane-bearing aqueous fluids from 273 to 523 K and from 1 to 2000 bar. Geochim. Cosmochim. Acta 70, 3369–3386.

Egger, M., Lenstra, W., Jong, D., Meysman, F.J., Sapart, C.J., van der Veen, C., Röckmann, T., Gonzalez, S., Slomp, C.P. 2016. Rapid sediment sccumulation results in high methane effluxes from coastal sediments. PloS ONE, 11(8), e0161609.

Ellwood, M.J., Schneider, L., Potts, J., Batley, G.E., Floyd, J., Maher, W.A., 2016. Volatile selenium fluxes from selenium-contaminated sediments in an Australian coastal lake. Environ. Chem. 13, 68–75.

EPA, 2011. Global Anthropogenic Non-CO 2 Greenhouse Gas Emissions : 1990 - 2030, Office of Atmospheric Programs Climate Change Division U.S. Environmental Protection Agency. NW Washington, DC.

Espinosa-Ortiz, E., Rene, E.R., van Hullebusch, E.D., Lens, P.N.L., 2015. Removal of selenite From wastewater in a *Phanerochaete chrysosporium* pellet based fungal bioreactor. Int. Biodeterior. Biodegrad. 102, 361–369.

Fei, Q., Guarnieri, M.T., Tao, L., Laurens, L.M.L., Dowe, N., Pienkos, P.T., 2014. Bioconversion of natural gas to liquid fuel: Opportunities and challenges. Biotechnol. Adv. 32, 596–614.

Fermoso, F.G., Collins, G., Bartacek, J., O'Flaherty, V., Lens, P.N.L., 2008. Acidification of Methanol-Fed Anaerobic Granular Sludge Bioreactors by Cobalt Deprivation: Induction and Microbial Community Dynamics. Biotechnol. Bioeng. 99, 49–58

Haynes, C. a, Gonzalez, R., 2014. Rethinking biological activation of methane and conversion to liquid fuels. Nat. Chem. Biol. 10, 331–339.

Hockin, S.L., Gadd, G.M., 2003. Linked redox precipitation of sulfur and selenium under anaerobic conditions by sulfate-reducing bacterial biofilms. Appl. Enivironmental Microbiol. 69, 7063–7072.

Howarth, A.J., Katz, M.J., Wang, T.C., Platero-prats, A.E., Chapman, K.W., Hupp, J.T., Farha, O.K., 2015. High efficiency adsorption and removal of selenate and selenite from water using metal – organic frameworks. J. Am. Chem. Soc. 137(23), 7488-7494.

Jain, R., Matassa, S., Singh, S., van Hullebusch, E.D., Esposito, G., Lens, P.N.L., 2016. Reduction of selenite to elemental selenium nanoparticles by activated sludge. Environ. Sci. Pollut. Res. 23, 1193–1202.

Jain, R., Seder-Colomina, M., Jordan, N., Dessi, P., Cosmidis, J., van Hullebusch, E.D., Weiss, S., Farges, F., Lens, P.N.L., 2015. Entrapped elemental selenium nanoparticles affect physicochemical properties of selenium fed activated sludge. J. Hazard. Mater. 295, 193–200.

Jin, T., Yeol, E., 2016. Metabolic versatility of microbial methane oxidation for biocatalytic methane conversion. J. Ind. Eng. Chem. 35, 8–13.

Lai, C.-Y., Wen, L.-L., Shi, L.-D., Zhao, K.-K., Wang, Y.-Q., Yang, X., Rittmann, B.E., Zhou, C., Tang, Y., Zheng, P., Zhao, H.-P., 2016. Selenate and nitrate bioreductions using methane as the electron donor in a membrane biofilm reactor. Environ. Sci. Technol. 50, 10179−10186.

Leahy, M., Barden, J.L., Murphy, B.T., Slater-thompson, N., Peterson, D., 2013. International Energy Outlook 2013.

Lemly, A.D., 2004. Aquatic selenium pollution is a global environmental safety issue. Ecotoxicol. Environ. Saf. 59, 44–56.

Lenz, M., Hullebusch, E.D. Van, Hommes, G., Corvini, P.F.X., Lens, P.N.L., 2008. Selenate removal in methanogenic and sulfate-reducing upflow anaerobic sludge bed reactors. Water Res. 42, 2184–2194.

Liang, L., Guan, X., Huang, Y., Ma, J., Sun, X., Qiao, J., 2015. Efficient selenate removal by zero-valent iron in the presence of weak magnetic field 156, 1064–1072.

Losi, M.E., Frankenberger, W.T., Valley, S.J., 1997. Reduction of selenium oxyanions by *Enterobacter cloacae* SLD1a-1 : Isolation and growth of the bacterium and its expulsion of selenium particles 63, 3079–3084.

Lunsford, J.H., 2000. Catalytic conversion of methane to more useful chemicals and fuels: A challenge for the 21st century. Catal. Today 63, 165–174.

Macaskie, L.E., Mikheenko, I.P., Yong, P., Deplanche, K., Murray, A.J., Paterson-Beedle, M., Coker, V.S., Pearce, C.I., Cutting, R., Pattrick, R.A.D., Vaughan, D., van der Laan, G., Lloyd, J.R., 2010. Today's wastes, tomorrow's materials for environmental protection. Hydrometallurgy 104, 483–487.

Macfarquhar MJK, Broussard DL, Burk RF, Dunn JR, Green AL (2011) Acute selenium toxicity associated with a dietary supplement. Arch. Int. Med. 170(3), 256–261.

Mal, J., Nancharaiah, Y.V, Hullebusch, E.D. Van, Lens, P.N.L., 2017. Biological removal of selenate and ammonium by activated sludge in a sequencing batch reactor. Bioresour. Technol. 229, 11–19.

Mal, J., Nancharaiah, Y.V., van Hullebusch, E.D., Lens, P.N.L., 2016. Effect of heavy metal co-contaminants on selenite bioreduction by anaerobic granular sludge. Bioresour. Technol. 206, 1–8.

Mcglynn, S.E., Chadwick, G.L., Kempes, C.P., Orphan, V.J., 2015. Single cell activity reveals direct electron transfer in methanotrophic consortia. Nature 526, 531–535.

Mueller, T.J., Grisewood, M.J., Nazem-Bokaee, H., Gopalakrishnan, S., Ferry, J.G., Wood, T.K., Maranas, C.D., 2015. Methane oxidation by anaerobic archaea for conversion to liquid fuels. J. Ind. Microbiol. Biotechnol. 42, 391–401.

Muscatello, J.R., Belknap, A.M., Janz, D.M., 2008. Accumulation of selenium in aquatic systems downstream of a uranium mining operation in northern Saskatchewan, Canada. Environ. Pollut. 156, 387–393.

Nancharaiah, Y.V, Lens, P.N.L., 2015. Ecology and biotechnology of selenium-respiring bacteria. Microbiol. Mol. Biol. Rev. 79, 61–80.

Nancharaiah, Y.V, Lens, P.N.L., 2015. Selenium biomineralization for biotechnological applications. Trends Biotechnol. 33, 323–330.

Nazem-Bokaee, H., Gopalakrishnan, S., Ferry, J.G., Wood, T.K., Maranas, C.D., 2016. Assessing methanotrophy and carbon fixation for biofuel production by Methanosarcina acetivorans. Microb. Cell Fact. 15, 10.

Serra, M.C.C., Pessoa, F.L.P., Palavra, A.M.F., 2006. Solubility of methane in water and in a medium for the cultivation of methanotrophs bacteria. J. Chem. Thermodyn. 38, 1629–1633.

Soo, V.W.C., McAnulty, M.J., Tripathi, A., Zhu, F., Zhang, L., Hatzakis, E., Smith, P.B., Agrawal, S., Nazem-Bokaee, H., Gopalakrishnan, S., Salis, H.M., Ferry, J.G., Maranas, C.D., Patterson, A.D., Wood, T.K., 2016. Reversing methanogenesis to capture methane for liquid biofuel precursors. Microb. Cell Fact. 15, 11.

Subedi, G., Taylor, J., Hatam, I., Baldwin, S.A., 2017. Simultaneous selenate reduction and denitri fi cation by a consortium of enriched mine site bacteria. Chemosphere 183, 536–545.

Tan, L.C., Nancharaiah, Y.V, van Hullebusch, E.D., Lens, P.N.L., 2016. Selenium: Environmental significance, pollution, and biological treatment technologies. Biotechnol. Adv. 34, 886–907.

Tan, L.C., Nancharaiah, Y.V, van Hullebusch, E.D., Lens, P.N.L., 2016. Selenium: Environmental significance, pollution, and biological treatment technologies. Biotechnol. Adv. 34, 886–907.

Tang, C., Huang, Y.H., Zeng, H., Zhang, Z., 2014. Reductive removal of selenate by zero-valent iron : The roles of aqueous Fe(2+) and corrosion products, and selenate removal mechanisms. Water Res. 67, 166–174.

Tinggi, U., 2003. Essentiality and toxicity of selenium and its status in Australia : a review. Toxicol. Lett. 137, 103–110.

U.S. E.I.A., 2010. International Energy Statistics.

Viamajala, S., Bereded-Samuel, Y., Apel, W.A., Petersen, J.N., 2006. Selenite reduction by a denitrifying culture: Batch- and packed-bed reactor studies. Appl. Microbiol. Biotechnol. 71, 953–962.

Yang, L., Ge, X., Wan, C., Yu, F., Li, Y., 2014. Progress and perspectives in converting biogas to transportation fuels. Renew. Sustain. Energy Rev. 40, 1133–52.

Zehr JP, Oremland RS. 1987. Reduction of Selenate to Selenide by Sulfate-Respiring Bacteria: Experiments with Cell Suspensions and Estuarine Sediments. Appl. Environ. Microbiol. 53(6), 1365-69.

Zhang, Y., Henriet, J.P., Bursens, J., Boon, N., 2010. Stimulation of in vitro anaerobic oxidation of methane rate in a continuous high-pressure bioreactor. Bioresour. Technol. 101, 3132–38.

Zimmerman, M.T., Bayse, C.A., Ramoutar, R.R., Brumaghim, J.L., 2015. Sulfur and selenium antioxidants : Challenging radical scavenging mechanisms and developing structure – activity relationships based on metal binding. J. Inorg. Biochem. 145, 30–40.

CHAPTER 7

Volatile fatty acid production from Kraft mill foul condensate in upflow anaerobic sludge blanket reactors

This chapter will be modified and published as:

Eregowda, T., Kokko, M.E., Rene, E.R., Rintala, J., Lens, P.N.L., 2019. VFA production from Kraft mill foul condensate in upflow anaerobic sludge blanket (UASB) reactors. Manuscript submitted to Journal of Hazardous Materials.

Abstract

The utilization of foul condensate (FC) collected from a Kraft pulp mill for the anaerobic production of volatile fatty acids (VFA) was tested in upflow anaerobic sludge blanket (UASB) reactors operated at 22, 37 and 55 °C at a hydraulic retention time (HRT) of ~ 75 h. The FC consisted mainly of 11370, 500 and 592 mg/L methanol, ethanol and acetone, respectively. 42-46% of the organic carbon (methanol, ethanol and acetone) was utilized in the UASB reactors operated at an organic loading of ~ 8.6 gCOD/L.d and 52-70% of the utilized organic carbon was converted into VFA. Along with acetate, also propionate, isobutyrate, butyrate, isovalerate and valerate were produced from the FC. Prior to acetogenesis of FC, enrichment of the acetogenic biomass was carried out in the UASB reactors for 113 d by applying operational parameters that inhibit methanogenesis and induce acetogenesis. Activity tests after 158 d of reactor operation showed that the biomass from the 55 °C UASB reactor exhibited the highest activity after the FC feed compared to the biomass from the reactors at 22 and 37 °C. Activity tests at 37 °C to compare FC utilization for CH_4 versus VFA production showed that an organic carbon utilization > 98% for CH_4 production occurred in batch bottles, whereas the VFA production batch bottles showed 51.6% utilization. Furthermore, higher concentrations of C_3-C_5 VFA were produced when FC was the substrate compared to synthetic methanol rich wastewater.

Keywords: Foul condensate; UASB; acetogenesis; VFA production; Kraft pulp mill

7.1. Introduction

Condensates are generated during the chemical recovery from black liquor in the chemical pulping process of the pulp and paper industry and have varying concentrations of organic compounds (Eregowda et al., 2018; Rintala and Puhakka, 1994). The Kraft mill evaporator condensates or Kraft condensates account for up to 40% of the biological oxygen demand (BOD) but only 5% of the total wastewater flow generated (Meyer and Edwards, 2014b; Rintala and Puhakka, 1994). The methanol concentration in the Kraft condensates vary from 1-46 g/L and contributes to ~ 80-96% of the total chemical oxygen demand (COD) (Badshah et al., 2012; Dufresne et al., 2001). Along with methanol, the Kraft condensates also contain ethanol, turpentine, organic reduced sulfur compounds like methyl mercaptan, dimethyl sulfide and dimethyl disulfide, small amounts of ketones, terpenes, phenolics, resin and fatty acids (Meyer and Edwards, 2014b; Suhr et al., 2015).

Steam stripping and condensation are the most commonly used technologies for the treatment of Kraft condensates to reuse water and reduce the organic load to the wastewater treatment plant, after which the stripped non-condensable gases (NCG) rich in volatile organic compounds are incinerated (Suhr et al., 2015). Considering the high COD content (mostly as methanol), Kraft condensates hold a huge potential for biological resource recovery through the production of valuable end-products. Anaerobic biomineralization of methanol is mainly carried out through either direct conversion of methanol to CH_4 by methanogens or indirectly through intermediate formation of acetate by methylotrophic acetogens, followed by further conversion of acetate to CH_4 by acetoclastic methanogens (Lu et al., 2015; Paula L Paulo et al., 2004; Weijma and Stams, 2001). Few studies have reported the anaerobic digestion of Kraft condensates in upflow anaerobic sludge blanket (UASB) reactors for biogas production (Dufresne et al., 2001; Liao et al., 2010; Minami et al., 1986; Xie et al., 2010) and a lab-scale efficiency of up to 86% COD removal has been achieved (Badshah et al., 2012).

Waste-derived volatile fatty acids (VFA), especially acetate, produced from waste and side stream biomass are valuable bioconversion products (Lee et al., 2014). They can be applied in industries such as polymer, food and fertilizer, and as precursors to fuels and chemicals (Eregowda et al., 2018; Lee et al., 2014). No study has so far reported the production of VFA through the acetogenesis of foul condensates in batch or continuous reactor systems. During acetogenesis, methanol is utilized by obligate anaerobes such as methylotrophic bacteria and homoacetogens that synthesize acetyl Co-A from methanol, resulting in acetate as the end-product. Acetogenesis occurs only in the presence of more oxidised carbon containing compounds such as formate, CO and CO_2 (Florencio, 1994).

The objectives of this study were, therefore: (i) to enrich the acetogens in UASB reactors, (ii) to examine the acetogenesis of FC, (iii) to study the effect of temperature (22, 37 and 55 °C) on the enrichment of acetogens and acetogenesis of FC, and (iv) to compare the effect of enrichment of acetogens and FC feed on the reactor biomass activity.

7.2. Materials and methods

7.2.1. Source of biomass and media composition

The anaerobic granular sludge used as the inoculum was collected from a mesophilic UASB reactor treating wastewater from the integrated production of the beta-amylase enzyme and ethanol from oat (Jokioinen, Finland). The synthetic wastewater used for the enrichment of the acetogens consisted of 1.2 g NaCl, 0.4 $MgCl_2 \cdot 6H_2O$, 0.5 g KCl, 0.3 g NH_4Cl, 0.15 g $CaCl_2$ and 0.2 g KH_2PO_4 per 1000 ml of MilliQ water. The cobalt (Co) deprived trace element solution consisted of (in mg/L): 1500 $FeCl_2 \cdot 4H_2O$, 100 $MnCl_2 \cdot 4H_2O$, 70 ZnCl, 62 H_3BO_3, 36 $Na_2MoO_4 \cdot 2H_2O$, 24 $NiCl_2 \cdot 6H_2O$ and 17 $CuCl_2 \cdot 2H_2O$.

7.2.2. Characteristics of the foul condensate (FC)

The FC, one of the Kraft condensate streams with a high concentration of methanol and total reduced sulfur (TRS), was collected from a chemical pulp mill in Finland and stored in airtight plastic containers in a fume hood prior to the experiments. The characteristics of the FC (***Table 7.1***) were as follows: pH (9.3 ± 0.2), conductivity (1374 µS/m), TRS (236 ± 25 mg/L) and organic compounds, namely, acetone (592 ± 49 mg/L), methanol (11370 ± 688 mg/L), ethanol (500 ± 26 mg/L), acetate (60 ± 2.5 mg/L) and propionate (30.5 ± 2.3 mg/L). The FC was supplemented with 1 mL/L Co deprived trace element solution and 0.2 g/L yeast extract and the pH was adjusted to 7.5 (± 0.2).

7.2.3. Experimental setup

Three laboratory scale UASB reactors (***Figure 7.1***) with a working volume of 0.5 L (height = 40 cm, inner diameter = 2.2 cm) were used to study the anaerobic utilization of methanol at room temperature (22 ± 2 °C, R22), mesophilic (37 ± 3 °C, R37) and thermophilic (55 ± 3 °C, R55) conditions at a hydraulic retention time (HRT) of 12 h in phases M1-M7, 24 h in M8 and 75 h in M9, respectively. The temperatures of the reactors R37 and R55 were maintained using heated water jackets. The reactors were inoculated with ~ 150 mL anaerobic granular sludge (36 ± 4 g/L volatile suspended solids) and the influent was fed from the bottom of the reactor with an upflow velocity of 0.033 m/h. The gas produced from the UASB reactors were collected in air-tight gas collection bags.

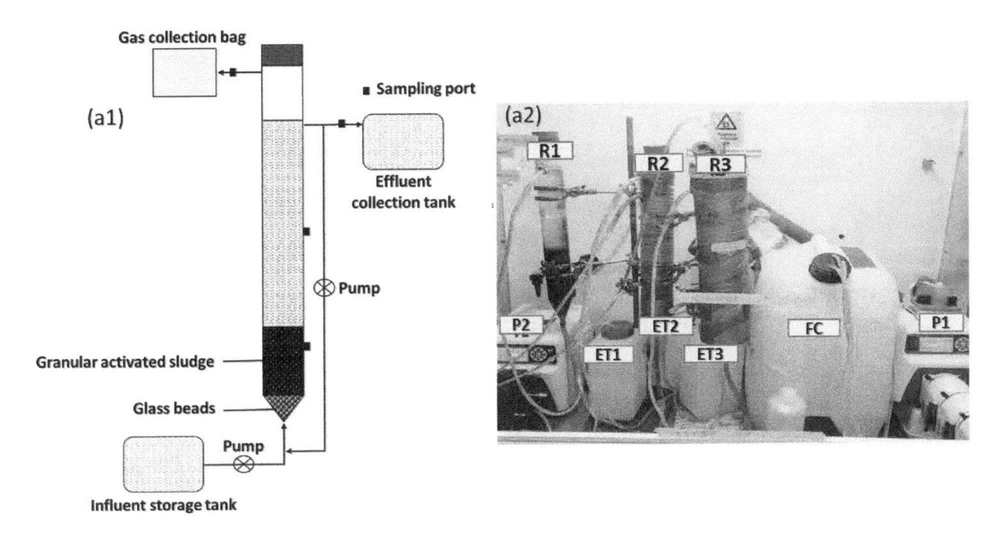

Figure 7.1: (a1) Schematic of the upflow anaerobic sludge blanket (UASB) reactor for the treatment of the Kraft mill foul condensate. The recirculation loop (in blue color) was incorporated with a recirculation ratio of 3 during the phase F1 operation when the UASB reactors were fed with FC. (a2) Picture of the UASB reactors during phase F1 operation with the FC feed. R22, R37, R55 are the UASB reactors operated at 22, 37 and 55 °C, respectively, and ET1, ET2 and ET3 were their respective effluent tanks. P1 and P2 were the feed and recirculation pumps, respectively. FC is the feed foul condensate holding tank.

7.2.4. Operational phases of the UASB reactors

The operation of the three UASB reactors, R22, R37 and R55 was carried out in nine phases (***Table** 7.2*). During the first six phases (M1-M6), methanogenic activity was detected and different operational strategies were tested to inhibit the methanogens. In phase M1, synthetic methanol-rich wastewater (SMWW) was fed to the UASB reactors at an OLR of ~ 1 g/L.d. During phase M2, 1 mL/L of Co deprived trace element solution was incorporated in the SMWW. The SMWW pH was reduced to 4.0 in phase M3, whereas in phase M4, the pH was restored to 6.5 and 0.33 g/L of HCO_3^- was added to the SMWW. The OLR was increased from 1 g/L.d to 2 g/L.d in phase M5. The influent pH was decreased from 6.5 to 5.5 in phase M6. In phase M7, 1 g/L of 2-bromoethanesulfonate (BES) was added to the SMWW. The HRT was increased from 12 h to 24 h in phase M8 to enhance the methanol utilization. Phase M9 was operated by feeding the reactors with FC. To avoid a concentration gradient along the height of the UASB reactors, a recirculation loop was incorporated at a recirculation ratio of 3 during phase M9.

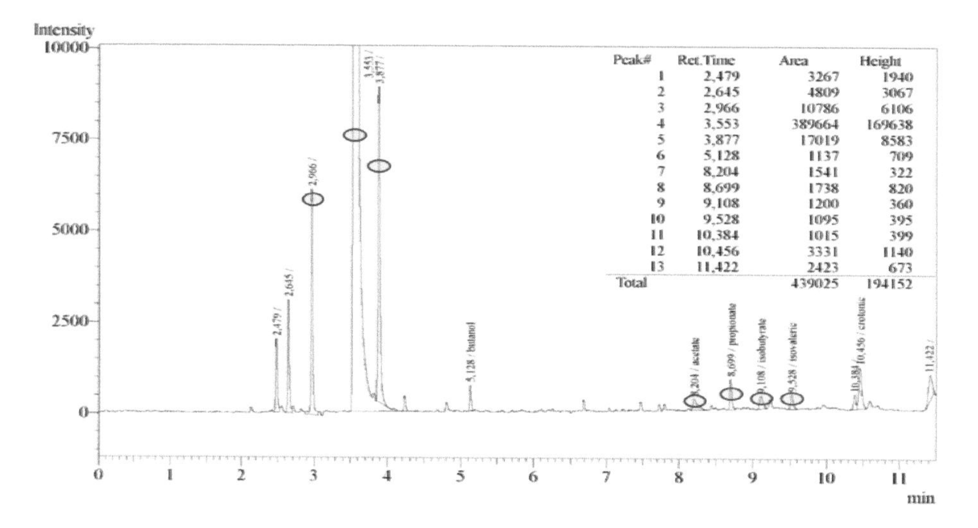

Figure 7.2: *GC-FID analysis of the FC with a run time of 11 min. The peaks of the organic compounds that were characterized are indicated with a circle.*

7.2.5. Batch activity tests

To compare the UASB reactor biomass activity for methanogenic and acetogenic methanol utilization, three sets of batch tests (T1, T2 and T3) were carried out. The tests were performed in duplicate, in 500 mL (for T1) and 120 mL (for T2 and T3) airtight serum bottles having a working volume of 300 and 60 mL, respectively. The biomass inoculum size was ~ 5 g/L VS for all the batch incubations. T1 batch tests were aimed at comparing the organic carbon utilization between synthetic media and FC for CH_4 and VFA production. They were inoculated with 100 mL fresh sludge and incubated at 37 °C and consisted of: (a) SMWW (10 g/L methanol), (b) SMWW + BES (1 g/L), (c) FC (200 mL, undiluted), (d) FC + BES (1 g/L). The T2 and T3 batch tests were inoculated with 20 mL sludge collected on day 113 (prior to feeding FC) and on day 158 (after feeding FC), respectively, from the reactors R22, R37 and R55 and incubated at 22, 37 and 55 °C along with 40 mL undiluted FC as the substrate.

Table 7.1: *Composition of different types of kraft condensates and COD removal values reported in the literature.*

Type of condensate (bioprocess)	pH	Total COD (mg/L)	Methanol (mg/L)	Sulfide (mg/L)	Other compounds (mg/L)	COD removal (%)	Reference
Foul condensate (aerobic; treatment in membrane bioreactor)		5120	3893	-	TRS (345)	> 90	Dias et al., (2005)
Methanol condensate (anaerobic; CH_4 production in UASB reactors)	9.6	1095 ± 22	600 ± 2	Na	Ammonium (21.1) Sulfate (410)	84-86	Badshah et al. (2012)
Combined Kraft condensates (anaerobic; CH_4 production in UASB reactors)	nm	700-4000	1300	210	Nm	59-90	Meyer and Edwards (2014)
Kraft evaporator condensates (anaerobic; CH_4 production in UASB and submerged anaerobic membrane bioreactors)	nm	600-6500	375-2500	1-690	Ethanol (0-190); 2-propanol (0-18); Acetone (1.5-5.1); Phenols (17-42); Terpenes (0.1-660); Sulfite (3-10) Resin acids (28-230)	70-99	Meyer and Edwards (2014) Xie et al. (2010)
Contaminated evaporator condensate (anaerobic; CH_4 production in column bioreactors)	9.5	1500	1000	604	Sulfate (12)	> 95	Dufresne et al. (2001)
Foul condensate (anaerobic; VFA production in UASB reactors)	9.3 ± 0.2	25700 ± 2650	11370 ± 688		Ethanol (500 ± 26) Acetone (592 ± 49) TRS (236 ± 25) Acetate (60 ± 2.5) Propionate (30.5 ± 2.3)	42-46	This study

7.2.6. Analytical methods

Liquid samples from the UASB reactor and batch incubations were filtered using 0.45 μm syringe filters and stored at 4 °C before analysing the influent and effluent pH, methanol and VFA concentrations. Additionally, for phase M9 of the reactor operation and for batch incubations (T1, T2 and T3) with FC as the substrate, the influent and effluent ethanol and acetone concentrations were analysed. The gas samples were collected in air-tight syringes and analysed immediately for methane content and the gas volume was recorded by measuring the water displacement with an air-tight water column. All the analysis were carried out in duplicates and the data presented in the plots are the average values with a standard deviation < 1.0%.

The CH_4 concentration was analysed using a gas chromatograph (Shimadzu GCe2014, Japan) equipped with a Porapak N column (80/100 mesh), a thermal conductivity detector (TCD) with N_2 as the carrier gas. The temperature of the oven, injector and detector were 80, 110 and 110 °C, respectively. The concentrations of VFA (acetate, propionate, isobutyrate, butyrate, isovalerate and valerate), methanol, acetone and ethanol were analysed using a gas chromatograph fitted with a flame ionization detector (Shimadzu GC-2010, Japan) (Kinnunen et al., 2015). The TRS concentration of the FC was measured spectrophotometrically (DR/2010, HACH, Loveland, CO) using a HACH sulfide kit (HACH, USA) after appropriate dilution.

Phases M1-M6 were dominated by methanogenesis and enrichment of acetogens was achieved starting from phase M7. The methanol utilization, CH_4 production and VFA production profiles are shown in ***Figures 7.3a, 7.3b, and 7.4***, respectively. The operational parameters in different phases of the UASB reactor operation, methanol utilization and the carbon recovery are shown in ***Table 7.2***. For the mass balance, organic carbon utilization was calculated as the fraction of methanol (additionally ethanol and acetone when FC was the substrate) utilized to the fraction of methanol supplied. The organic carbon recovery was calculated as the stoichiometric fraction of methanol (additionally ethanol and acetone when FC was the substrate) converted to VFA or CH_4 to the fraction of methanol utilized.

Table 7.2: *Summary of the operational parameters, average organic carbon utilization and average organic carbon recovery during different phases of UASB reactor operation.*

Phase	Operational period (d)	Parameters	Influent pH	COD (g/L)	OLR (g/L.d)	HRT (h)	Average organic carbon utilization (%)*			Average organic carbon recovery (%)+		
							R22	R37	R55	R22	R37	R55
M1	1-3	Methanol rich synthetic wastewater (pH:6.5)	6.5	0.5	1.1	12	10.8	19.8	16.5	0.3	5	1.7
M2	4-8	Addition of Co deprived trace element solution	6.5	0.5	1.1	12	10.8	95	28.4	6.4	51	8.2
M3	9-11	Low pH shock at pH: 4.0	4.0	0.5	1.1	12	22.7	63.8	16.3	0	11.1	1.72
M4	12-27	Carbonate addition	6.5	0.5	1.1	12	26.8	88.4	78.9	8.5	65.4	40.2
M5	28-37	OLR doubled	6.5	1	2.2	12	44.5	97.8	88.6	16.6	64.2	58.0
M6	38-55	pH: 5.5	5.5	1	2.2	12	32.3	77.4	25.1	17.5	41.5	8.4
M7	56-92	2-bromoethanesulfonate	5.5	1	2.2	12	28.7	23.7	19.5	90.6	59.6	89
M8	93-113	HRT doubled	6.5	1	1.1	24	53.3	53	49.7	108.8	126.4	128.1
M9	114-158	Foul condensate feed	7.5	26.9	8.6	75	42.3	45.9	45.8	57.1	52	69.5

Note: * Total carbon utilization of phase M9 comprises of both acetone and ethanol utilization.

+Stoichiometric fraction of methanol converted to CH_4 in phases M1-M6 or VFA in phases M7-M9, to the fraction of methanol utilized (additionally acetone and ethanol in the phase M9).

Table 7.3: *Summary of the organic carbon utilized and recovered as VFA in batch test T1 that compares CH_4 vs VFA production from synthetic wastewater vs foul condensate*

	Synthetic media		Foul condensate	
	- BES	+ BES	- BES	+ BES
Organic carbon utilization (%)*	99.8	72	98.7	51.6
Organic carbon utilization rate (mg/L.d)	503.5	202.8	191.8	110.3
Organic carbon recovery as VFA (%)+	1.8	96.1	8.5	78.5
Maximum VFA production (mg/L)	61.8	5120	299	2880
Organic carbon recovery as CH_4 (%)	76.4	na	60.1	Na
Maximum methane production (mL)	1151.5	na	638	Na

7.3. Results

7.3.1. Enrichment of acetogens in the UASB reactors

7.3.3.1. Methanogenic phase

To improve the methanol utilization efficiency, which was < 25% during the first three days (phase M1) of operation (***Figure 7.3a***), a Co deprived trace element solution was incorporated in the influent on day 4 (phase M2). The methanol utilization efficiency in reactor R37 quickly improved from < 20% to > 92% within 1.5 d, whereas in the reactors R22 and R55, the maximum methanol utilization efficiencies were 11 and 32%, respectively. The CH_4 production increased from < 2, 20 and 5 mL/d to 40, 134 and 18 mL/d in the reactors R22, R37 and R55, respectively, within 2 days.

The reactors were subjected to a pH shock, wherein the influent pH was adjusted to 4.0 (phase M3, days 9-11) to inhibit methanogenesis. A rapid decrease in the CH_4 production was observed in all three reactors, i.e to values < 13 mL/d. The methanol utilization also dropped from 57, 95 and 30% to 5, 42 and 4%, respectively, in the reactors R22, R37 and R55 due to the pH shock. In phase M4 (days 12-27), when the pH was restored to 6.5 and bicarbonate was incorporated in the influent as a co-substrate (4 moles of methanol and 2 moles of bicarbonate are required to produce 3 moles of acetate), the methanol utilization increased rapidly to > 95% in the reactors R37 and R55, respectively, and to 30% in the reactor R22. Up to 10.2, 82 and 61% of the consumed methanol was recovered as CH_4 in the reactors R22, R37 and R55, respectively, during the phase M4 (***Figure 7.3b***).

In the phase M5, when the OLR was doubled (from 1 to 2 g/L.d), the methanol utilization dropped to 24, 86, 73% in the reactors R22, R37 and R55, respectively. However, these values increased to ~ 84.5, 98 and 97% within 3 d. Furthermore, with a brief drop in CH_4 production at the beginning of phase M5, methanogenesis gradually increased and up to 30, 76, and 61% of the methanol was recovered as CH_4 in reactors R22, R37 and R55, respectively, on day 36. The influent pH was reduced from 6.5 to 5.5 in M6 to enhance VFA accumulation and inhibit CH_4 production. The methanol utilization dropped to ~ 21.5, 76 and 17.5% and the CH_4 production dropped to 13.2, 25.2 and 2.1% in reactors R22, R37 and R55, respectively, on day 42. However, both methanol utilization and CH_4 production gradually increased in the three UASB reactors. During the entire methanogenic period (phase M1-M6), acetate (~ 50 mg/L) was observed occasionally in the effluent. Since the concentrations of VFA produced in all the three UASB reactors were insignificant, VFA was not considered in the mass balance for the phases M1-M6.

Table 7.4: *Summary of the organic carbon utilized and conversion to VFA in batch activity tests to compare the effect of FC feed on the biomass of the UASB reactors (Batch test T2 and T3)*

	Before feeding the foul condensate			**After feeding the foul condensate**		
	22 °C	37 °C	55 °C	22 °C	37 °C	55 °C
Organic carbon utilization (%)*	68.7	72	67.3	41.8	50.7	56.9
Organic carbon utilization rate (mg/L.d)	168.4	176.4	165	102.4	124.3	139.2
Organic carbon recovery as VFA (%)+	95	92.7	92.3	95.2	98.3	91.9
Maximum VFA production (mg/L)	3993	4069	3857	2336	3028	3141

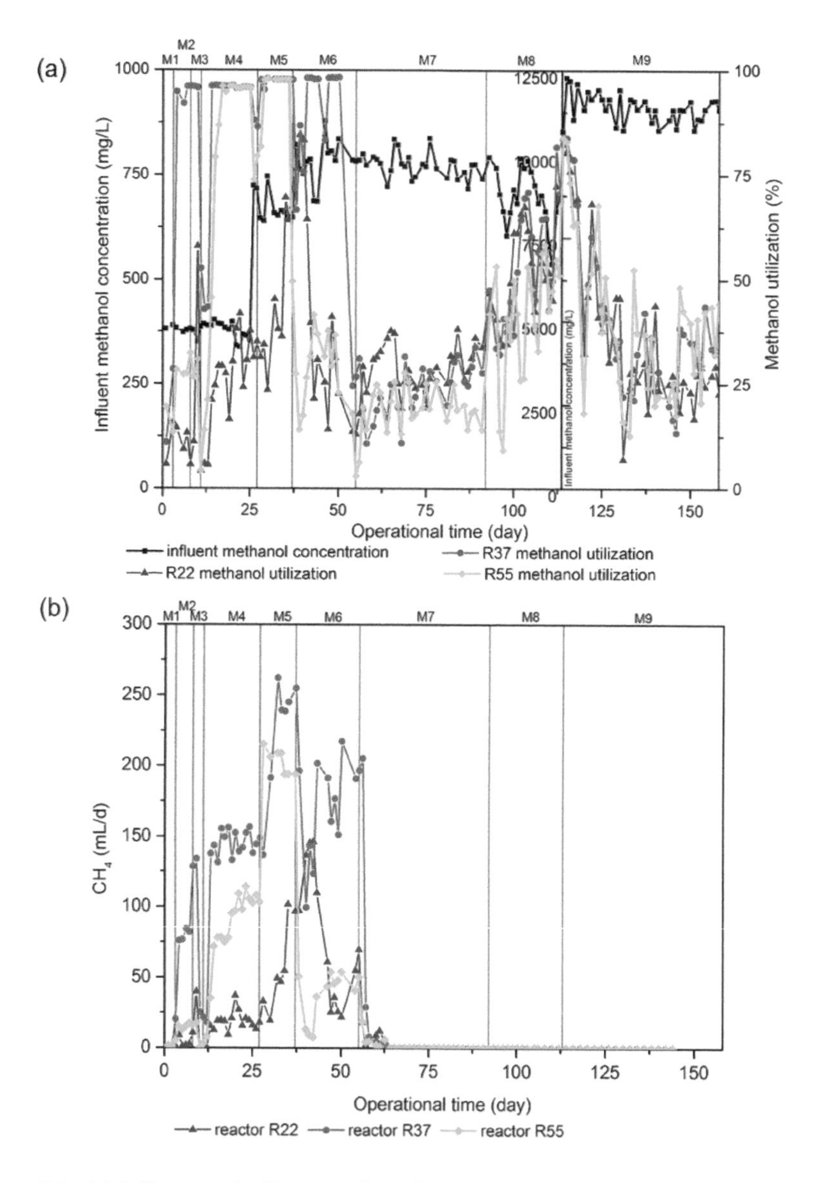

***Figure 7.3**: (a) Influent and effluent methanol concentration profiles at different temperatures and (b) Methane concentration profiles during different operational phases. M1-M6 are the methanogen dominant phases, M7-M8 are the acetogen dominant phases and M9 is the FC feeding phase.*

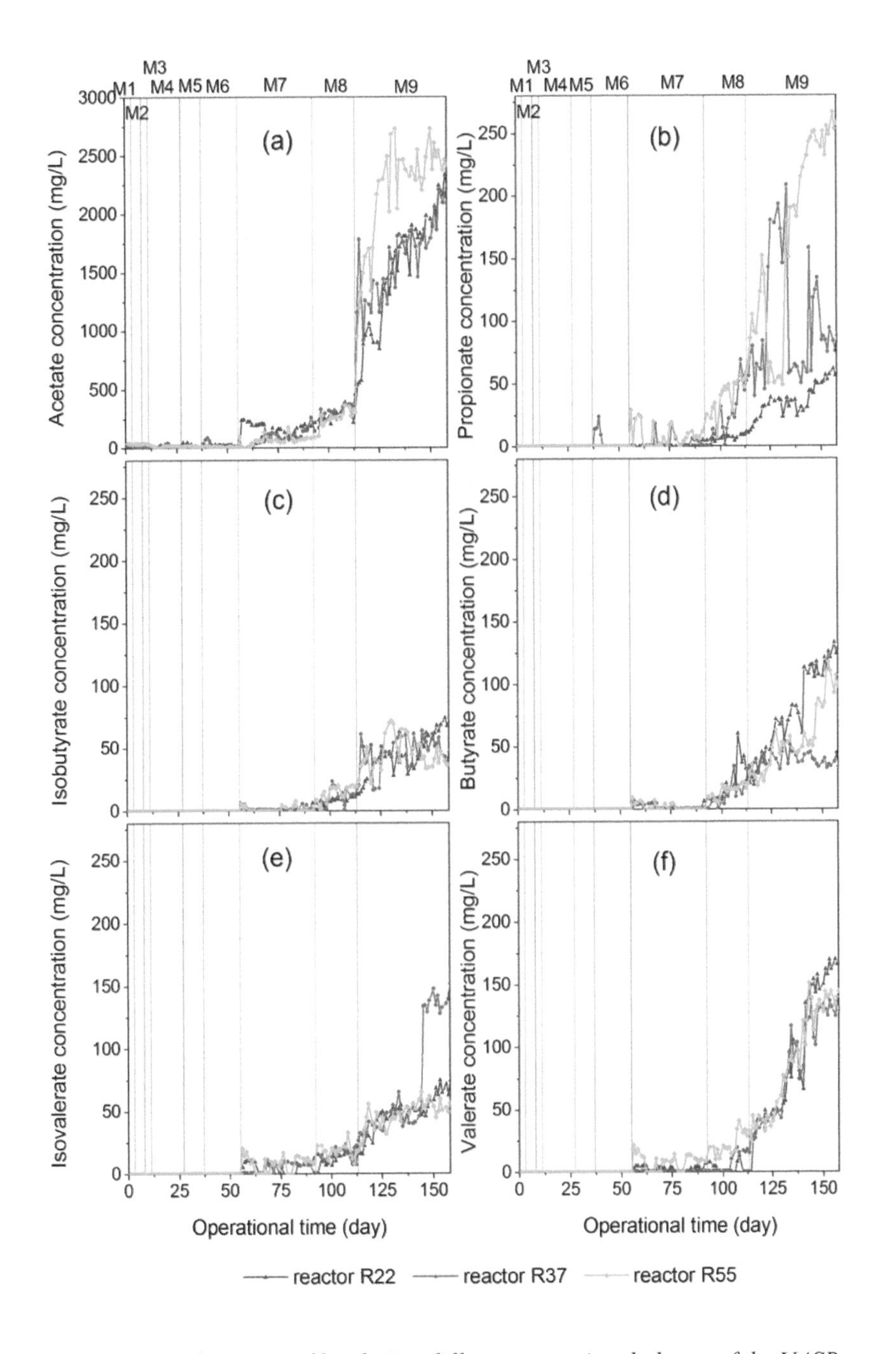

Figure 7.4: *VFA production profiles during different operational phases of the UASB reactors R22, R37 and R55. M1-M6 are the methanogen dominant phase, M7-M8 are the acetogen dominant phases and M9 is the FC feeding phase.*

7.3.1.2. Acetogenic phase

In phase M7, the addition of BES (1 g/L) completely inhibited the methanogenesis and CH_4 production in the three reactors was insignificant (< 10 mL/d). A maximum methanol utilization of up to 28.7, 23.7and 19.5% was achieved (***Table 7.2***) in the reactors R22, R37 and R55 along with a production of up to 244, 120 and 75 mg/L acetate, respectively. Additionally, up to ~ 9 and 6 mg/L of respectively, isovalerate and valerate in the reactor R22; 25, 6 and 9 mg/L of respectively, propionate, isobutyrate and isovalerate in the reactor R37 and ~ 20, 17 and 14 mg/L of respectively, propionate, isovalerate, and valerate in the reactor R55 were produced on day 91.

In order to increase the methanol utilization rate for VFA production, the HRT was increased from 12 h to 24 h in phase M8. Consequently, the methanol utilization increased to 53.3, 53 and 49.7% and the acetate production increased to 392, 377, 356 mg/L in reactors R22, R37 and R55, respectively, on day 95. Furthermore, ~ 10, 56 and 50 mg/L propionate, ~ 12, 14 and 18 mg/L isobutyrate, ~ 20, 15 and 18 mg/L butyrate, ~ 17, 19 and 22 mg/L isovalerate were, respectively, produced in the reactors R22, R37 and R55 and ~ 30 mg/L isovalerate was produced in R55. Overall, the percentage methanol utilization was lower, i.e. ~ 45.6, 43.6 and 52.4% during the acetogenic period (phases M7 and M8) compared to the methanogenic period (phases M1-M6) where the utilization was up to 84.4, 97.8 and 88.6% in the reactors R22, R37 and R55, respectively (***Figure 7.3a)***.

7.3.2. Acetogenesis of the foul condensate (FC)

An average methanol removal efficiency of 36.4, 40 and 41.3% was observed in the reactors R22, R37 and R55, respectively, in phase M9 (FC feeding phase). During the 44 d operation of phase M9, a steady utilization of 84, 79.2 and 82.2% acetone, and 85.7, 86.5 and 86.2% ethanol was, respectively, observed in the reactors R22, R37 and R55 (***Figure 7.5***). The individual VFA production was as follows: Acetate- 1623, 1660 and 2274 mg/L; propionate- 39.5, 100, 178 mg/L; isobutyrate- 47.6, 47.1, 46.5 mg/L; butyrate- 86, 39.5 and 61.2 mg/L; isovalerate- 51.6, 79 and 49.7 mg/L and valerate- 105.3, 91.8 and 98 mg/L in the reactors R22, R37 and R55, respectively. The CH_4 production in the three UASB reactors was insignificant (< 10 mL/d). The organic carbon utilization and the carbon recovery as VFA during phase M9 were, respectively, 42.3 and 57.1% in R22, 45.9 and 52% in R37, and 45.8 and 69.5% in R55 (***Table 7.2***).

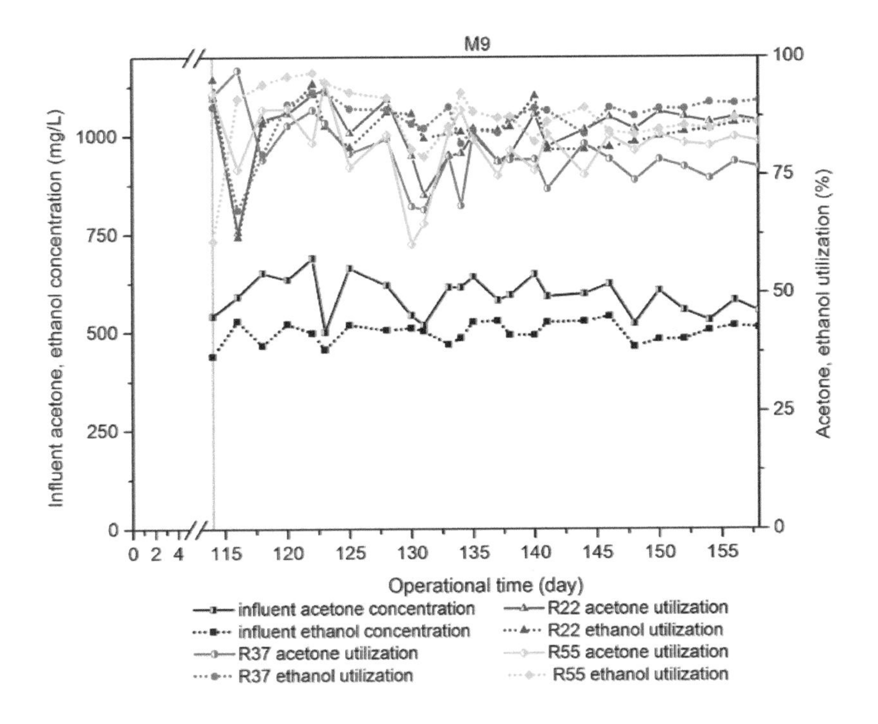

Figure 7.5: *Influent and effluent acetone (solid lines) and ethanol (dotted lines) concentration profiles during the FC feeding phase (M9) of the UASB reactors R22, R37 and R55.*

7.3.3. Batch activity tests

7.3.3.1. Batch study with the inoculum (T1)

The profiles of methanol utilization, VFA production and CH_4 production from FC in batch study T1 are shown in ***Figures 7.6, 7.7 and 7.8***, respectively. The methanol utilization for CH_4 production in batch incubations without BES (98.7%) was higher compared to VFA production batch incubations (51.6%) when FC was used as the substrate. A similar trend was observed with SMWW as the substrate, wherein > 98% methanol utilization occurred in the batch incubations without BES and 72% in the batch incubations with BES (***Table 7.3***). In batch incubations without BES, nearly complete methanol utilization occurred within 18 d when SMWW was the substrate and the organic carbon utilization required ~ 48 d when FC was the substrate.

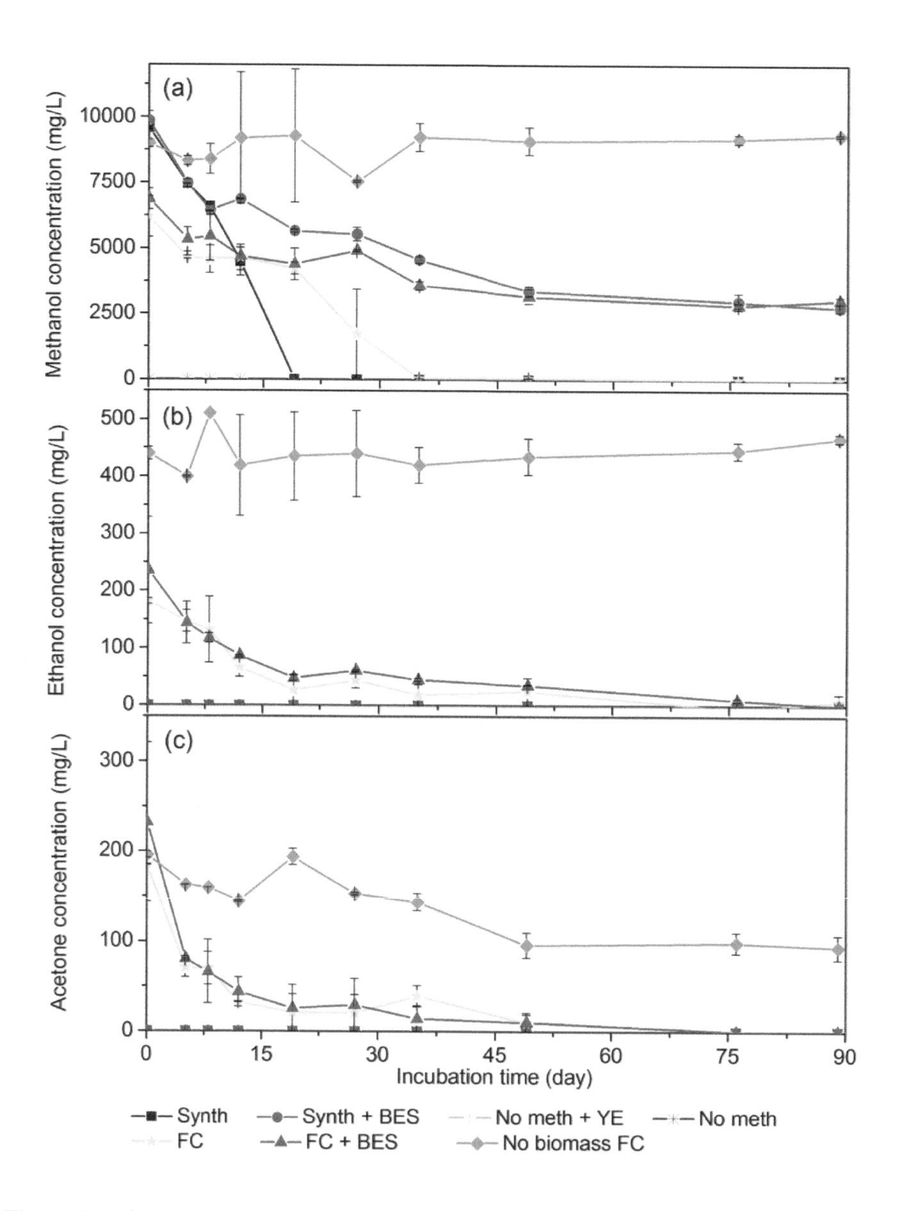

Figure 7.6*: (a) Methanol, (b) Ethanol and (c) Acetone concentration profiles during the batch test T1 with FC or SMWW as substrate. 'synth' and 'FC' are, respectively, the batches with SMWW and FC as the substrate, 'synth+BES' and 'FC+BES' are the batches with SMWW and FC, along with the addition of BES, 'nometh+YE' are the batches with yeast extract in SWW without methanol, 'no biomass FC' is the negative control with only FC and no biomass, and 'no meth' is the negative control with only biomass in milli-Q water.*

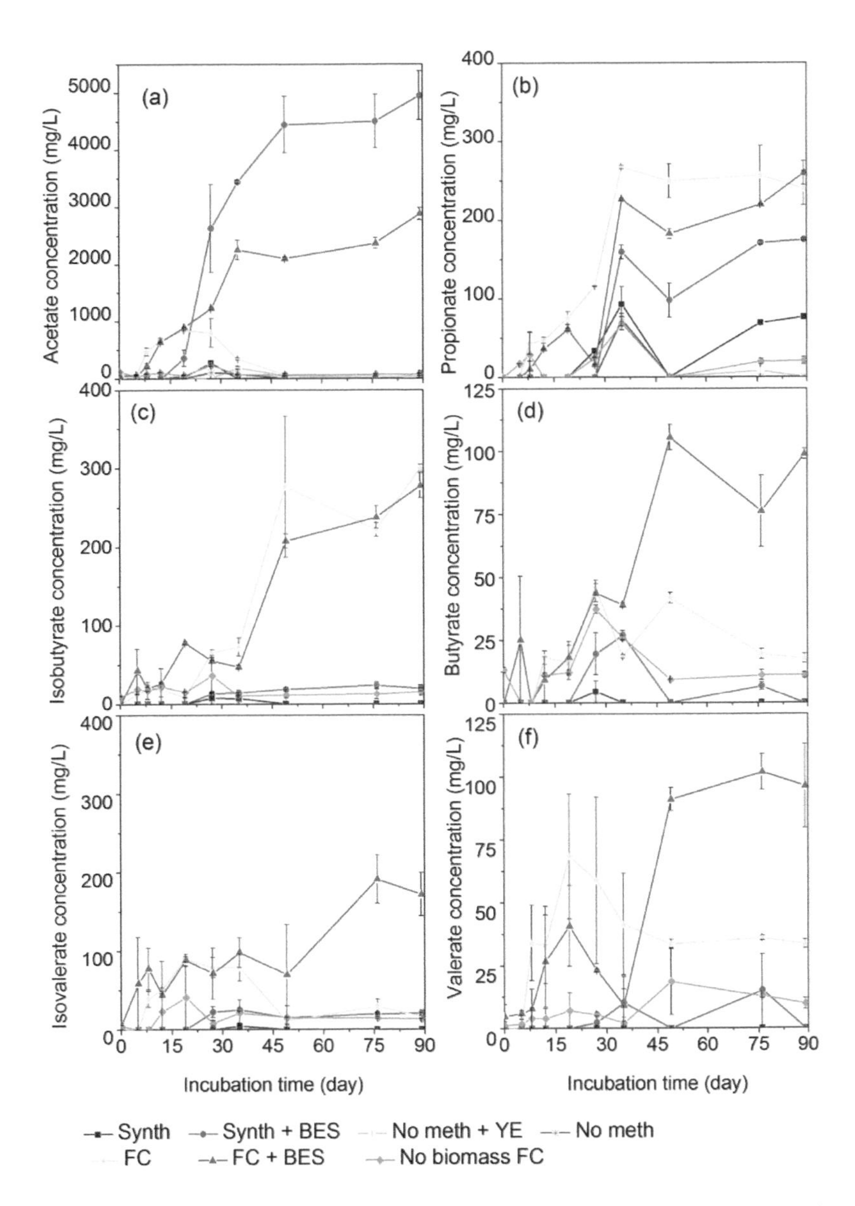

***Figure 7.7**: (a) Acetate, (b) Propionate, (c) Isobutyrate, (d) Butyrate, (e) Isovalerate and (f) Valerate concentration profiles during the batch test T1 fed with FC or SMWW. 'synth' and 'FC' are, respectively, the batches with SMWW and FC as the substrate, 'synth+BES' and 'FC+BES' are the batches with SMWW and FC, along with the addition of BES, 'nometh+YE' are the batches with yeast extract in SWW without methanol, 'no biomass FC' is the negative control with only FC and no biomass, and 'no meth' is the negative control with only biomass in milli-Q water.*

Concerning the VFA production, ~ 900 mg/L acetate was produced on day 15 and was eventually consumed (in 45 d) in the batch incubations with FC as the substrate and without BES. Up to 260 mg/L propionate, 294 mg/L isobutyrate, 25 mg/L butyrate, 21 mg/L isovalerate and 36 mg/L valerate were also produced. In the batch incubations containing BES, up to 3560 mg/L acetate, 255 mg/L propionate, 255 mg/L isobutyrate, 100 mg/L butyrate, 175 mg/L isovalerate and 90 mg/L valerate were produced from the FC. Although the percentage methanol utilization was higher when SMWW was the substrate, only acetate (~ 4950 mg/L) and propionate (150 mg/L) production were significant and the concentration of isobutyrate, butyrate, isovalerate and valerate were < 25 mg/L (***Figure 7.7***). Furthermore, from the batch incubations with only biomass or only yeast extract (no methanol or FC), VFA production was insignificant, implying that the biomass or yeast extract did not contribute to the VFA production.

7.3.3.2. Batch study using UASB biomass after 113 (T2) and 158 (T3) days operation

The profiles of methanol, ethanol and acetone utilization and VFA production for the batch studies T2 and T3 are shown in ***Figures 7.9 and 7.10***, respectively. The organic carbon utilization was 68.7, 72, 67.3% for the biomass from reactors R22, R37 and R55, respectively, with an organic carbon recovery of 95, 92.7 and 92.3% (***Table 7.4***) for the batch study T2 (biomass collected after acetogen enrichment and before feeding FC feed). The activity of the biomass was clearly lower in batch study T3 (biomass collected after feeding FC) since the methanol utilization decreased to 41.8, 50.7 and 56.9% for the biomass from reactors R22, R37 and R55, respectively. The organic carbon converted to VFA was 95.2, 98.3 and 91.9%, respectively, indicating that acetogenesis was the main pathway, both before and after feeding the FC to the sludge sampled from the three UASB reactors.

The organic carbon utilization of the unenriched biomass (51.6%), enriched biomass (72%) and the biomass after the FC feed (50.7%) for the batch incubations at 37 °C shows that the organic carbon utilization increased after 113 d of enrichment, but decreased upon feeding FC. Reactor R55 was the least affected by the FC feed since the organic carbon utilization before and after feeding FC was, respectively, 67.3 and 56.9% (***Table 7.4***).

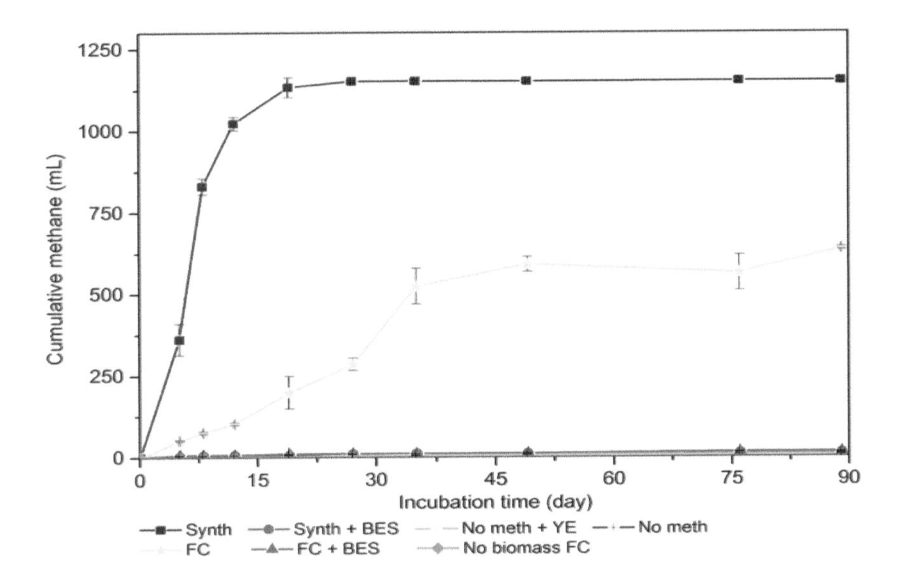

***Figure 7.8**: Cumulative methane production during the batch test T1 fed with FC or SMWW. 'synth' and 'FC' are the batches with SMWW and FC as the substrate, 'synth' and 'FC' are, respectively, the batches with SMWW and FC as the substrate, 'synth+BES' and 'FC+BES' are the batches with SMWW and FC, along with the addition of BES, 'nometh+YE' are the batches with yeast extract in SWW without methanol, 'no biomass FC' is the negative control with only FC and no biomass, and 'no meth' is the negative control with only biomass in milli-Q water.*

7.4. Discussion

7.4.1. Foul condensate (FC) utilization in the UASB reactors

This study showed that VFA can be produced by acetogenesis of real FC from a chemical pulp industry in UASB reactors operated at 22, 37 and 55 °C (***Figure 7.4***). The organic compounds present in the FC (methanol, ethanol and acetone) served as the substrates for the acetogenic biomass for the VFA production. The average organic carbon utilization from FC for VFA production was, respectively, 36.6, 40 and 42.7% in reactors R22, R37 and R55 (***Table 7.2, Figure 7.3a***). Although few studies on the biological treatment of condensates report a COD removal > 98% (***Table 7.1***), these studies applied aerobic treatment or anaerobic digestion for CH_4 production.

The limiting factors for the conversion of organic carbon to VFA from FC were likely: (a) unavailability of bicarbonate, resulting in the absence of buffering and co-substrate for the acetogenesis (discussed in section 4.3) and (b) the concentration of the TRS compounds present in the FC, which was as high as 236 mg/L. Acetogens have a higher tolerance for sulfide toxicity compared to methanogens (Chen et al., 2014). However, H_2S species can penetrate their cell membrane and suppress the cellular activity (Bundhoo and Mohee, 2016; Dhar et al., 2012). Furthermore, the presence of high concentrations of sulfide (~ 236 mg/L) could lead to the precipitation of trace elements as metal sulfides.

Since the FC is a condensate formed during the evaporation of black liquor in a chemical pulping mill (Tran and Vakkilainnen, 2012), it is as such devoid of any salts and complex micronutrient source that could provide the required vitamins, nitrogen source, nutritional (sodium, phosphorus and calcium) and trace (such as zinc, manganese, magnesium, copper, nickel and iron) elements necessary for cellular activity (Zhang et al., 2012). Although FC was supplemented with a Co deprived trace element solution and 200 mg/L yeast extract, the supplied vitamins and trace elements could have been precipitated by the high sulfide concentrations, thus influencing their bioavailability and limiting the bioconversions (Chen et al., 2014). Further studies to increase the methanol utilization in the UASB reactors under long term operation, by varying operational parameters such as HRT, OLR, and recirculation rate are necessary.

It is noteworthy to mention that the biological removal of ethanol and acetone in any Kraft condensate is reported for the first time in this study. The removal of ethanol (> 80%) and acetone (75-80%) was higher in the three reactors investigated compared to that of methanol (30-40%). Acetogens and other methylotrophic bacteria are able to readily assimilate ethanol and acetone as the carbon source (Florencio, 1994). Moreover, the concentrations of ethanol and acetone were much lower in the FC compared to methanol (***Table 7.1***).

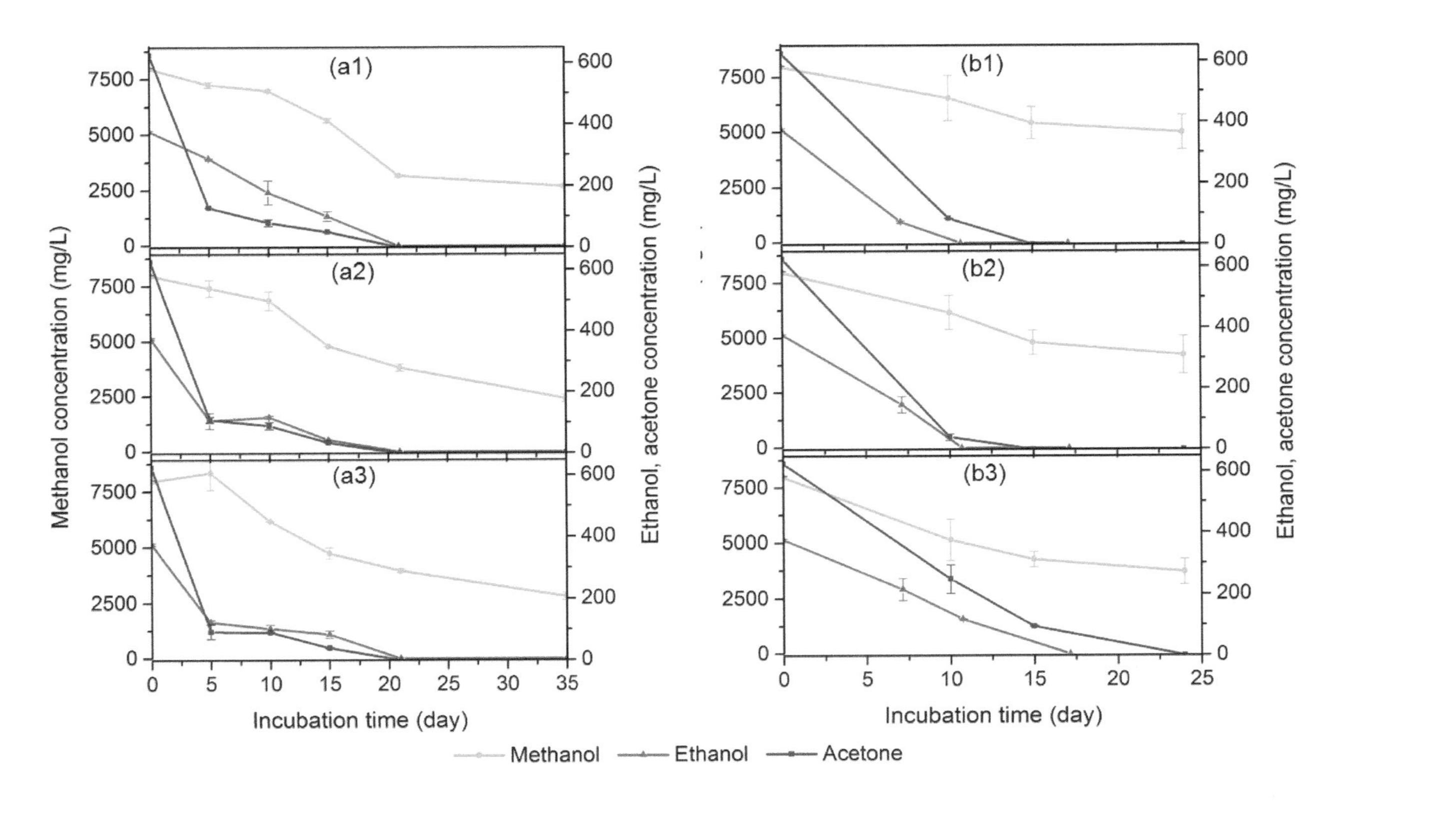

Figure 7.9: Methanol, ethanol and acetone concentration profiles in batch activity tests: a1, a2 and a3 are the tests performed with the biomass collected from the reactors R22, R37 and R55, respectively, before the FC feed (T2), while b1, b2 and b3 are the tests performed with biomass from the respective biorectors collected after the FC feed (T3).

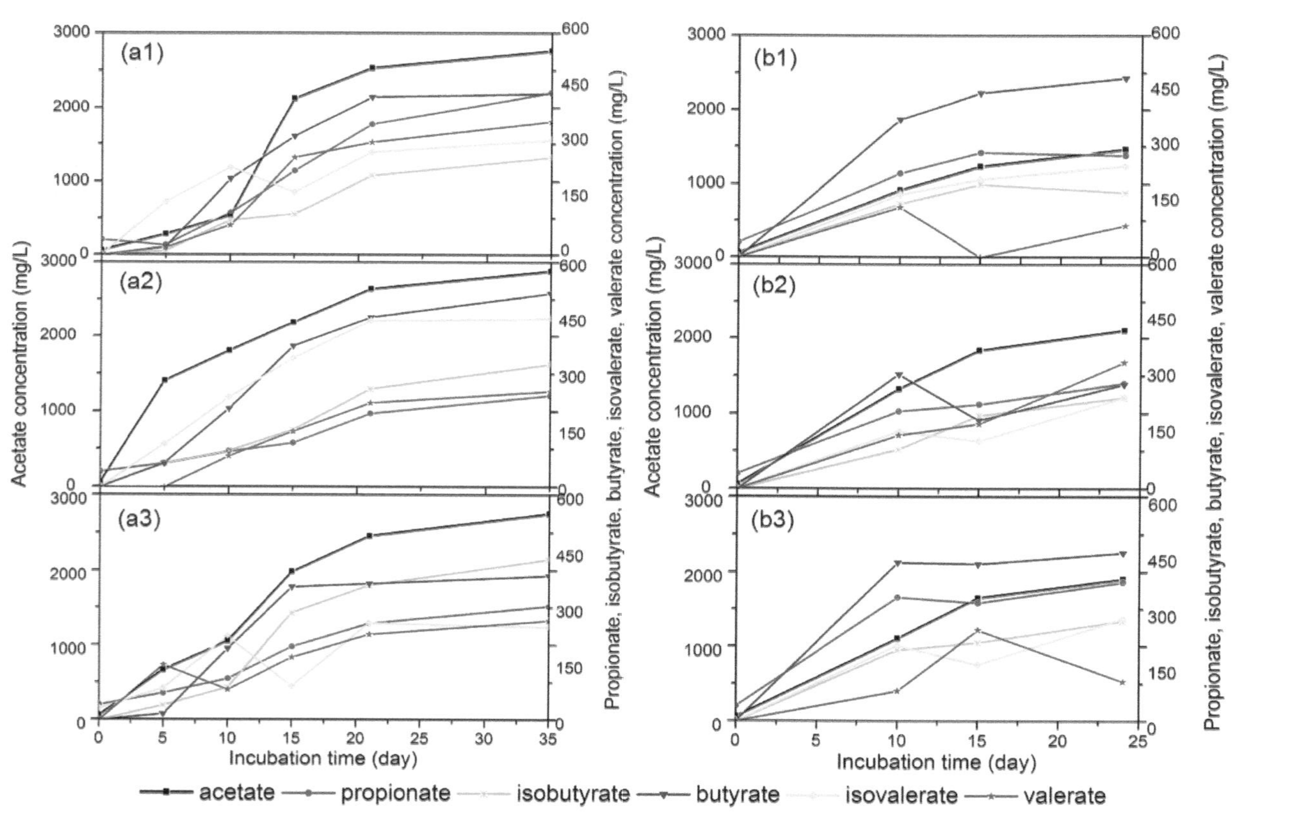

Figure 7.10: *VFA concentration profiles during the batch tests: : a1, a2 and a3 are the tests performed with the biomass collected from the reactors R22, R37 and R55, respectively, before the FC feed (T2), while b1, b2 and b3 are the tests performed with biomass collected from the respective bioreactors after the FC feed (T3).*

Enrichment of acetogens

The anaerobic mineralization of methanol mainly occurs through methanogens, resulting in CH_4 production and acetogens, resulting in acetate formation (Lin et al., 2008). Prior to feeding the reactors with FC, inhibition of methanogens and the enrichment of acetogens was tested (using SMWW) through a systematic change of previously reported process parameters such as pH in phase M3 and M6 (Costa and Leigh, 2014), the presence of bicarbonate in phase M4 (Florencio, 1994; Paula L. Paulo et al., 2004; Paulo et al., 2003), higher organic loading rates in phase M5 (Florencio, 1994; Weijma and Stams, 2001) and Co deprivation throughout the operational period (Fermoso et al., 2008; Florencio et al., 1994; Paula L Paulo et al., 2004). However, inhibition of methanogenesis was not successful until the application of BES in phase M7 (1 g/L) (Braga et al., 2016). Although BES is a well-established specific inhibitor of methanogens (Zhuang et al., 2012), often used in the lab-scale reactors for hydrogen production through dark fermentation or bioelectrochemical processes (Braga et al., 2016), it was used as the last option in this study (phase M7-M9) to avoid operational costs over additional residual COD to the influent.

During the methanogen dominant phases (M1-M6), methanol was most likely utilized by the hydrogenotrophic methanogens that are Co independent (Fermoso et al., 2008; Paula L. Paulo et al., 2004). In phase M6, the bicarbonate addition induced buffering and additionally, the activity of the hydrogenotrophic methanogens (1 mole of CH_4 is produced using 1 mole of bicarbonate and 4 moles of hydrogen) promoted the recovery of the methanogenic activity (Paulo et al., 2003; Weijma and Stams, 2001), especially in the reactors R37 and R55 (***Figure 7.3, Table 7.2***). The high methanogenic activity can also be attributed to the fact that the anaerobic granular sludge used as the inoculum for the UASB reactors was collected from a mesophilic UASB reactor treating the wastewater from the integrated production of beta-amylase enzyme and ethanol from oat and a large methanogenic bacterial population was thus already present in the inoculum.

7.4.2. Biomass activity and CH_4 vs VFA production

Comparing the methanogen inhibited batch incubations in T1, up to 72% methanol utilization was achieved with SMWW and 51.6% when FC was the substrate. The unavailability of bicarbonate was likely the main limiting factor for the lower utilization of methanol in the FC. This could be confirmed from the fact that the organic carbon recovery as VFA (%) in the methanogen inhibited SMWW batch incubations was 96.1% compared to 78.5% for the FC batch incubations (***Table** 7.3*).

Under the tested conditions, CH_4 production (> 98%) resulted in higher organic carbon utilization than VFA production (~ 51.6%). Few studies have reported CH_4 production from Kraft condensate in reactor configurations such as UASB, submerged membrane bioreactors and column bioreactors. The COD removal efficiencies achieved in these studies are described in ***Table** 7.1*. Concerning VFA, the acidogenic fermentation of food waste resulted in the production of up to 6.3 g/L VFA at an organic loading of 15 kg COD/m^3.day (Dahiya et al., 2015). In this study, production of up to 2274, 178, 46, 61, 49 and 98 mg/L acetate, propionate, isobutyrate, butyrate, isovalerate and valerate were, respectively, achieved at an organic loading of 8.6 g COD/L.

Apart from VFA production from the FC constitutes, production of propionate, isobutyrate, butyrate, isovalerate and valerate in the three UASB reactors might be via carboxylate chain elongation (CCA). Comparing the VFA production profile in batch study T1, although the initial COD was similar, a higher amount of propionate, isobutyrate, butyrate, isovalerate and valerate were produced in the batches with FC as substrate compared to SMWW. The possibility of VFA production from other sources such as yeast extract and the degradation of the UASB sludge was eliminated from the T1 batch study with only yeast extract or biomass (without FC or SMWW as the substrate). Targeted VFA production in mixed culture studies through the CCA of acetate with ethanol or methanol as electron donors have been previously reported (Chen et al., 2017, 2016). For example, butyrate production through CCA of acetate with methanol or ethanol as electron donor (Coma et al., 2016; Roghair et al., 2018). Further investigations using ^{13}C labelled compounds are necessary to verify the effect of methanol, ethanol or acetone on VFA production via CCA.

7.4.3. Effect of temperature on UASB reactor performance

During the methanogenic period (phases M1-M6), reactor R37 showed the highest methanogenic activity and was the quickest to revert to methanogenesis upon changing the process parameters in each phase. Several studies on anaerobic digestion, and methanol utilization in particular, have shown that the operation of the bioreactors under mesophilic conditions showed the highest COD removal efficiency and CH_4 production (Berube and Hall, 2000; Chen et al., 2014; Rintala, 2011).

The influence of temperature on the VFA production is ambiguous since the optimal temperature is shown to be > 40 °C (Lu and Ahring, 2005; Mengmeng et al., 2009) in some studies, whereas few studies suggest mesophilic range (Zhang et al., 2009) conditions. The methanogenic (phases M1-M6) as well as acetogenic (phases M7-M8) activity were the least in reactor R22 compared to R37 and R55 during the enrichment phase (***Figure 7.3***). Furthermore, the operational temperature of the UASB reactor seemed to affect the production of individual VFA: R55 showed the largest production of acetate and propionate, while R37 showed the largest production of isovalerate and R22 showed the largest production of isobutyrate, butyrate and valerate (***Figure 7.4***).

Batch tests (T2 and T3) showed that the biomass activity was higher in the reactor R55 compared to the biomass sampled from R22 and R37 after the FC feed. This is advantageous for industrial situations since the temperature of FC in the chemical pulping process is in the range of 50-60 °C (Tran and Vakkilainnen, 2012). More research with adequate temperature control is required to examine the effect of temperature on the VFA production from Kraft condensates.

7.4.4. Practical implications

FC is a good candidate for VFA production due to the presence of high concentrations of methanol along with small amounts of ethanol and acetone. Owing to its composition, FC could also be used as the electron donor for biological treatment of wastewaters that are deprived of carbon source, e.g. biological nitrogen removal from municipal wastewater, treatment of mining and other sulfate rich wastewaters and effluents from the fertilizer industry.

Due to the industrial application of VFA, several studies have examined the production of waste-derived VFA from a wide range of waste organics and effluents such as activated sludge, primary sludge, food waste, kitchen waste, organic fraction of municipal solid waste, wood mill, paper mill, food processing industrial effluents like olive oil mill, palm oil mill, cheese whey, dairy sugar industry and wine (Lee et al., 2014; Zhang and Angelidaki, 2015). VFA recovery could be done as a stand-alone process, wherein the UASB reactor effluent is directly fed to downstream VFA recovery units. Several physicochemical processes such as electrodialysis, adsorption on ion-exchange resins, fractional distillation, solvent extraction and precipitation can be used for VFA recovery (Eregowda et al., 2018; Zhang and Angelidaki, 2015). VFA production from FC is a novel resource recovery approach for managing the condensates of a chemical pulping process, which are otherwise evaporated to incinerate the VOC and TRS rich non-condensable gas fraction.

7.5. Conclusions

VFA production from foul condensates was successfully demonstrated in UASB reactors operated at 22, 37 and 55 °C, with an average organic carbon utilization of 42.3, 45.9 and 45.8%, respectively. Prior to feeding FC, enrichment of acetogens was carried out. The biomass from the reactor R55 showed a higher activity after the FC feeding phase compared to reactors R22 and R37. Batch activity tests showed that CH_4 production led to complete utilization of FC compared to only 54.6% utilization in case of VFA production and the production of C_3-C_5 VFA was higher when FC was the substrate, compared to SMWW.

Acknowledgement

The authors thank Dr. Pritha Chatterjee (TUT, Tampere) for the analytical support during UASB operation, Prof. Tero Joronen (TUT, Tampere) for making the arrangements to procure the foul condensate. This work was supported by the Marie Skłodowska-Curie European Joint Doctorate (EJD) in Advanced Biological Waste-to-Energy Technologies (ABWET) funded from Horizon 2020 under the grant agreement no. 643071.

References

B.Q. Liao, K. Xie, H.J. Lin, D. Bertoldo, Treatment of kraft evaporator condensate using a thermophilic submerged anaerobic membrane bioreactor, Water Sci. Technol. 61 (2010) 2177–2183.

B.R. Dhar, E. Elbeshbishy, G. Nakhla, Influence of iron on sulfide inhibition in dark biohydrogen fermentation, Bioresour. Technol. 126 (2012) 123–130.

C. Mengmeng, C. Hong, Z. Qingliang, S.N. Shirley, R. Jie, Optimal production of polyhydroxyalkanoates (PHA) in activated sludge fed by volatile fatty acids (VFAs) generated from alkaline excess sludge fermentation, Bioresour. Technol. 100 (2009) 1399–1405.

F.G. Fermoso, G. Collins, J. Bartacek, V. O'Flaherty, P. Lens, Acidification of methanol-fed anaerobic granular sludge bioreactors by cobalt deprivation: induction and microbial community dynamics, Biotechnol. Bioeng. 99 (2008) 49–58.

H. Tran, E.K. Vakkilainnen, The Kraft chemical recovery process, TAPPI Kraft recovery process. (2012) 1–8.

J. Lu, B.K. Ahring, Effects of temperature and hydraulic retention time on thermophilic anaerobic pretreatment of sewage sludge, Conf. Proc. 1 (2005) 159–164.

J. Rintala, High-rate anaerobic treatment of industrial wastewaters, Water Sci. Technol. 24 (1991) 69–74.

J. Weijma, A.J.M. Stams, Methanol conversion in high-rate anaerobic reactors, Water Sci. Technol. 44 (2001) 7–14.

J.A. Rintala, J.A. Puhakka, Anaerobic treatment in pulp- and paper-mill waste management: A review, Bioresour. Technol. 47 (1994) 1–18.

J.K. Braga, L.A. Soares, F. Motteran, I.K. Sakamoto, M.B.A. Varesche, Effect of 2-bromoethanesulfonate on anaerobic consortium to enhance hydrogen production utilizing sugarcane bagasse, Int. J. Hydrog. Energy. 41 (2016) 22812–22823.

J.L. Chen, R. Ortiz, T.W.J. Steele, D.C. Stuckey, Toxicants inhibiting anaerobic digestion: A review, Biotechnol. Adv. 32 (2014) 1523–1534.

K. Minami, T. Horiyama, M. Tasaki, Y. Tanimoto, Methane production using a bio-reactor packed with pumice stone on an evaporator condensate of a kraft pulp mill, J. Ferment. Technol. 64 (1986) 523–532.

K. Xie, H.J. Lin, B. Mahendran, D.M. Bagley, K.T. Leung, S.N. Liss, B.Q. Liao, Performance and fouling characteristics of a submerged anaerobic membrane bioreactor for kraft evaporator condensate treatment, Environ. Technol. 31 (2010) 511–521.

K.C. Costa, J.A. Leigh, Metabolic versatility in methanogens, Curr. Opin. Biotechnol. 29 (2014) 70–5.

L. Florencio, J.I.M.A. Field, G. Lettinga, Importance of cobalt for individual trophic groups in anaerobic methanol-degrading consortium, Appl. Environ. Microbiol. 60 (1994) 227–234.

L. Florencio, The fate of methanol in anaerobic bioreactors, Thesis, Wageningen Agricultural University, Wageningen, The Netherlands.

L. Zhang, W. Ouyang, A. Li, Essential role of trace elements in continuous anaerobic digestion of food waste, 16 (2012) 102–111.

L. Zhuang, Q. Chen, S. Zhou, Y. Yuan, H. Yuan, Methanogenesis control using 2-bromoethanesulfonate for enhanced power recovery from sewage sludge in air-cathode microbial fuel cells, Int. J. Electrochem. Sci. 7 (2012) 6512–6523.

M. Badshah, W. Parawira, B. Mattiasson, Anaerobic treatment of methanol condensate from pulp mill compared with anaerobic treatment of methanol using mesophilic UASB reactors, Bioresour. Technol. 125 (2012) 318–327.

M. Coma, R. Vilchez-Vargas, H. Roume, R. Jauregui, D.H. Pieper, K. Rabaey, Product diversity linked to substrate usage in chain elongation by mixed-culture fermentation, Environ. Sci. Technol. 50 (2016) 6467–6476.

M. Roghair, T. Hoogstad, D.P.B.T.B. Strik, C.M. Plugge, P.H.A. Timmers, R.A. Weusthuis, M.E. Bruins, C.J.N. Buisman, Controlling ethanol use in chain elongation by CO_2 loading rate, Environ. Sci. Technol. 52 (2018) 1496–1506.

M. Suhr, G. Klein, I. Kourti, M.R. Gonzalo, G.G. Santonja, S. Roudier, L.D. Sancho, Best available techniques (BAT) - reference document for the production of pulp, paper and board, Eur. Comm. (2015) 1–906.

M.A.Z. Bundhoo, R. Mohee, Inhibition of dark fermentative bio-hydrogen production: A review, Int. J. Hydrogen Energy. 41 (2016) 6713–6733.

P. Zhang, Y. Chen, T.Y. Huang, Q. Zhou, Waste activated sludge hydrolysis and short-chain fatty acids accumulation in the presence of SDBS in semi-continuous flow reactors: Effect of solids retention time and temperature, Chem. Eng. J. 148 (2009) 348–353.

P.L. Paulo, B. Jiang, D. Cysneiros, A.J.M. Stams, Effect of cobalt on the anaerobic thermophilic conversion of methanol, Biotechnol. Bioeng. 85 (2004) 434-441.

P.L. Paulo, G. Villa, J.B. Van Lier, G. Lettinga, The anaerobic conversion of methanol under thermophilic conditions: pH and bicarbonate dependence, J. Biosci. Bioeng. 96 (2003) 213–218.

P.L. Paulo, M.V.G. Vallero, R.H.M. Treviño, G. Lettinga, P.N.L. Lens, Thermophilic (55°C) conversion of methanol in methanogenic-UASB reactors: influence of sulphate on methanol degradation and competition, J. Biotechnol. 111 (2004) 79–88.

P.R. Berube, E.R. Hall, Effects of elevated operating temperatures on methanol removal kinetics from synthetic kraft pulp mill condensate using a membrane bioreactor, Water Res. 34 (2000) 4359–4366.

R. Dufresne, A. Liard, M.S. Blum, Anaerobic treatment of condensates: trial at a kraft pulp and paper mill, Water Environ. Res. 73 (2001) 103–109.

S. Dahiya, O. Sarkar, Y. V Swamy, S.V. Mohan, Acidogenic fermentation of food waste for volatile fatty acid production with co-generation of biohydrogen, Bioresour. Technol. 182 (2015) 103–113.

T. Eregowda, L. Matanhike, E.R. Rene, P.N.L. Lens, Performance of a biotrickling filter for anaerobic utilization of gas-phase methanol coupled to thiosulphate reduction and resource recovery through volatile fatty acids production, Bioresour. Technol. 263 (2018) 591–600.

T. Meyer, E.A. Edwards, Anaerobic digestion of pulp and paper mill wastewater and sludge, Water Res. 65 (2014) 321–349.

V. Kinnunen, A. Ylä-Outinen, J. Rintala, Mesophilic anaerobic digestion of pulp and paper industry biosludge-long-term reactor performance and effects of thermal pretreatment, Water Res. 87 (2015) 105–111.

W.S. Chen, S. Huang, D.P.B.T.B. Strik, C.J.N. Buisman, Isobutyrate biosynthesis via methanol chain elongation: converting organic wastes to platform chemicals, J. Chem. Technol. Biotechnol. 92 (2017) 1370–1379.

W.S. Chen, Y. Ye, K.J.J. Steinbusch, D.P.B.T.B. Strik, C.J.N. Buisman, Methanol as an alternative electron donor in chain elongation for butyrate and caproate formation, Biomass and Bioenergy. 93 (2016) 201–208.

W.S. Lee, A.S.M. Chua, H.K. Yeoh, G.C. Ngoh, A review of the production and applications of waste-derived volatile fatty acids, Chem. Eng. J. 235 (2014) 83–99.

X. Lu, G. Zhen, M. Chen, K. Kubota, Y.Y. Li, Biocatalysis conversion of methanol to methane in an upflow anaerobic sludge blanket (UASB) reactor: Long-term performance and inherent deficiencies, Bioresour. Technol. 198 (2015) 691–700.

Y. Lin, Y. He, Z. Meng, S. Yang, Anaerobic treatment of wastewater containing methanol in upflow anaerobic sludge bed (UASB) reactor, Front. Environ. Sci. Eng. China. 2 (2008) 241–246.

Y. Zhang, I. Angelidaki, Bioelectrochemical recovery of waste-derived volatile fatty acids and production of hydrogen and alkali, Water Res. 81 (2015) 188–195.

CHAPTER 8

Volatile fatty acid adsorption on anion exchange resins: kinetics and selective recovery of acetic acid

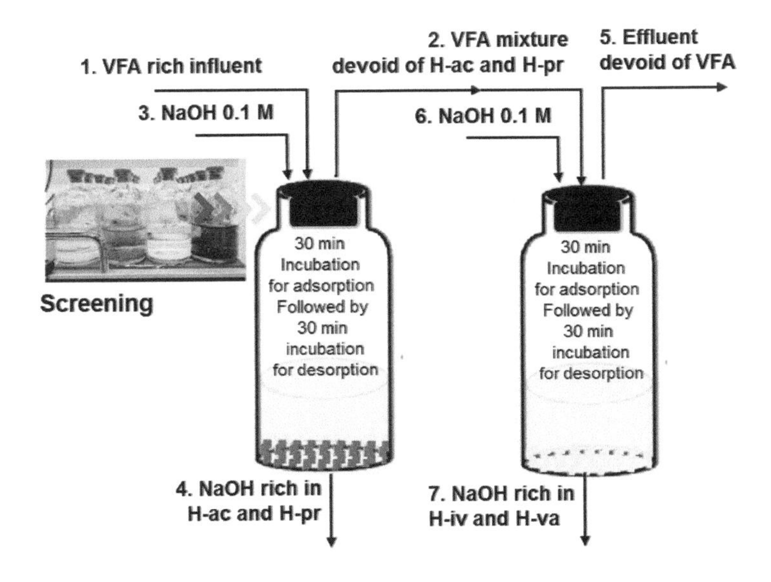

Modified version of this chapter was accepted for publication as:

Eregowda, T., Rene, E.R., Rintala, J., Lens, P.N.L., 2019. Volatile fatty acid adsorption on anion exchange resins: kinetics and selective recovery of acetic acid. Sep. Sci. Technol. 1-13.

Abstract

The removal of volatile fatty acids were examined through adsorption on anion exchange resins in batch systems. During the initial screening step, granular activated carbon and 11 anion exchange resins were tested and the resins Amberlite IRA-67 and Dowex optipore L-493 were chosen for further investigation. The adsorption kinetics and diffusion mechanism and adsorption isotherms of the two resins for VFA were evaluated. Based on the selective adsorption capacity of the resins, a sequential batch process was tested to achieve separation of acetic acid from the VFA mixture and selective recoveries >85% acetic acid and ~75% propionic acid was achieved.

Keywords: Volatile fatty acids; selective recovery; anion-exchange resins; Brunauer-Emmett-Teller model

8.1. Introduction

Volatile fatty acids (VFA) are used in a wide range of industries such as chemicals, polymers, food and agriculture. Most of these VFA are currently produced by the photo-catalysis of refined heavy oil and natural gas (Baumann and Westermann, 2016). In recent years, the recovery of VFA from the biological treatment of wastewater, sewage sludge, food and fermentation waste and lignocellulosic waste biomass through dark fermentation and anaerobic digestion processes have gained a lot of interest due to the economic value of the recovered VFA (Choudhari et al., 2015; Eregowda et al., 2018; Jones et al., 2017, 2015; Longo et al., 2014; Zhou et al., 2017). Furthermore, in anaerobic digesters and fermenters, the removal of excess VFA is beneficial for improving the process stability and performance because VFA are primary inhibitory intermediates in the process (Jones et al., 2015).

Several studies have reported that the recovery of individual VFA from bioreactors is more challenging than its production due to the similarities in their physico-chemical properties (Anasthas and Gaikar, 2001; Djas and Henczka, 2018; Martinez et al., 2015; Reyhanitash et al., 2016; Tugtas, 2011). Among the conventional physico-chemical recovery processes like electro-dialysis (Jones et al., 2015; Pan et al., 2018), fractional distillation, crystallization, precipitation, liquid-liquid extraction (Djas and Henczka, 2018), adsorption on ion exchange resins(Rebecchi et al., 2016; Tugtas, 2011; W.-L. Wu et al., 2016; Zhang and Angelidaki, 2015; Zhou et al., 2017) is a potential approach for the selective recovery of the desired VFA (López-garzón and Straathof, 2014). Although adsorption is generally based on the physical interaction between the adsorbent and the adsorbate, i.e. the VFA molecules, ion exchange is additionally based on chemisorption, wherein the formation of the ionic bonds between ionized carboxylic acid molecules and the cationic functional group of the anion exchange resin drives the process (López-garzón and Straathof, 2014). The performance of the ion exchange process for VFA recovery depends on the type of resin used, adsorption capacity of the resin, pH (Rebecchi et al., 2016), temperature (Da Silva and Miranda, 2013), type of desorption and the desorbing chemical used (Yousuf et al., 2016), presence of other competing anions (Reyhanitash et al., 2017) and the adsorption/contact time (Pradhan et al., 2017). In order to ameliorate the understanding of VFA recovery through adsorption, studies pertaining to several aspects (***Table 8.1***) have been reported in the literature.

Table 8.1: *Different aspects of VFA adsorption reported in different studies.*

Adsorption medium	Adsorbed compound	Adsorption resin	Key objective	Reference
Water	Acetic acid and glycolic acid	Amberlite IRA-67	Thermodynamic parameters and the effect of temperature	Uslu and Bayazit (2010)
A non-aqueous mixture of ethyl acetate and ethanol	Acetic acid	Indion 850, Tuision A-8X MP, Indion 810	Adsorption equilibrium	Anasthas and Gaikar (2001)
Artificial dark fermentation effluent containing chlorides, sulphates and phosphates and	Acetic, propionic, butyric and lactic acid	Lewatit VP OC 1065, Amberlite IRA96 RF, Amberlite IRA96 SB, and Lewatit VP OC 1064 MD PH	Adsorption from a complex solution mimicking fermented wastewater and desorption through nitrogen stripping at different temperatures for resin regeneration	Reyhanitash et al. (2017)
Binary systems using water, ethanol and n-propanol as solvents	Acetic, propionic and butyric acid	Purolite A133S and activated carbon	Adsorption and desorption using a multi-stage counter-current process	Da Silva and Miranda (2013)
Dark fermentation broth	Lactic, acetic and butyric acid	Amberlite IRA-67 and activated carbon	Adsorption and the effect of pH	Yousuf et al. (2016)
Acidogenic digestion effluent	Acetic, propionic, isobutyric, butyric, isovaleric, valeric,and caproic acid	Sepra NH2, Amberlyst A21, Sepra SAX, Sepra ZT-SAX	Adsorption and desorption using a basified ethanol and predicting the ion exchange capacity of the resin using an ion exchange model	Rebecchi et al. (2016)
Water	Acetic, propionic, isobutyric, butyric, isovaleric acid valeric acid	Amberlite IRA-67 and Dowex optipore L-493	Isotherm and kinetics of VFA adsorption and selective recovery using Amberlite IRA-67 and Dowex optipore L-493 in sequential batch mode	Chapter 8

However, there still exists a knowledge gap regarding the kinetics and mechanism of VFA adsorption on ion exchange resins. Furthermore, the separation of individual VFA using a combination of resins has not been reported in the literature. Therefore, batch adsorption experiments using single- and multi-component VFA were carried out to determine the kinetics and mechanism of VFA adsorption by the resins. Accordingly, the objectives of this study were: (i) to screen the anion exchange resins that are suitable for VFA adsorption, (ii) to determine the adsorption kinetics, mechanism and isotherm for the selected resins, and (iii) to examine the selective adsorption of individual VFA from a mixture of acetic (H-ac), propionic (H-pr), isobutyric (H-ib), butyric (H-bu), isovaleric (H-iv), valeric (H-va) acid using sequential batch adsorption and a combination of resins.

8.2. Materials and methods

8.2.1. Adsorption experiment

The properties (given by the manufacturers) of the 11 anion exchange resins and granular activated carbon (GAC) that were screened for the adsorption of VFA (H-ac, H-pr, H-ib, H-bu, H-iv and H-va) are listed in ***Table 8.2***. Batch adsorption experiments were carried out at the corresponding pK_a value of the aqueous VFA mixture, in 110 mL gastight serum bottles with a liquid volume of 50 mL without any resin pre-treatment. Experiments were performed in either duplicates or triplicates, at room temperature (22 ± 2 °C) on a rotary shaker at 200 rpm. The VFA concentration was monitored by collecting the liquid-phase samples, at regular time intervals as required for the individual adsorption tests. The collected samples were stored at 4 °C prior to analysis. The VFA concentrations were analysed using either a Shimadzu GC-2010 or Varian GC-430 gas chromatograph, both fitted with a flame ionization detector (FID) and a ZB-WAX plus (30 m × 0.25 mm) column or CP WAX-58 CB (30 m × 0.25 mm) column. Helium was used as the carrier gas. The oven temperature was maintained at 40 °C for 2 min, thereafter a ramp of 20 °C/min to reach 160 °C and another ramp of 40 °C/min to 220 °C and then maintained at 220 °C for 2 min.

8.2.2. Adsorption kinetics

The physical properties of the selected resins Amberlite IRA-67 (Amberlite) and Dowex Optipore L-493 (Dowex) are described in ***Table 8.3***. The adsorption kinetics of Amberlite and Dower was determined for a mixture containing 1 g/L each VFA (total: 6 g/L), at resin concentrations of 25, 50 and 75 g/L. The liquid samples were collected at intervals of 0, 2, 4,

6, 8 and 24 h from the batch bottles that were tested for 1 d and at intervals of 5, 10, 15, 20, 40, 60, 80, 100 and 120 min from the batch bottles that were tested for 2 h. The percentage adsorption and the equilibrium adsorption capacity were determined using Eq. 8.1 and Eq. 8.2:

$$\text{Percentage adsorption} = \frac{(C_0 - C_e) \times 100}{C_0} \qquad \text{(Eq. 8.1)}$$

$$\text{Equilibrium adsorption capacity (mg/g)}, \, q_e = \frac{(C_0 - C_e) \times V}{m} \qquad \text{(Eq. 8.2)}$$

To describe the kinetics of VFA adsorption, the pseudo-first-order, the pseudo-second-order and the Elovich's models were used. The differential and linear equations of the models are shown in Eq. 8.3–8.5; ***Table 8.4***, where, K_1 is the pseudo-first-order rate constant (min^{-1}), q_e is the amount of adsorbate in the adsorbent under equilibrium conditions (mg/g), t is the time of incubation (min), q_t is the adsorption at any given point of time (mg/g), K_2 is the pseudo-second-order rate constant (mg/g.min), h_0 is the initial adsorption rate (min^{-1}), β is the desorption constant (g/mg) and α is the initial adsorption rate (mg/g.min). Intraparticle diffusion model (Eq. 8.6; ***Table 8.4***) was used to examine the diffusion mechanism of the VFA molecules on the resin surface where, K_{id} (mg/g.$min^{-0.5}$) is intraparticle diffusion rate constant and C_{id} (mg/g) gives an approximation of the boundary layer thickness.

8.2.3. Resin regeneration

The regeneration of resin and the recovery of VFA through desorption was studied at different initial concentrations of NaOH (45, 60, 70, 80, 100 mM). After adsorption, the resins were subjected to treatment with NaOH for 2 h and samples were collected once every 20 min for 2 h.

8.2.4. Adsorption isotherm

The VFA adsorption isotherm for the single-component and multi-component systems were evaluated using the equilibrium adsorption capacity values (Eq. 8.2) for Amberlite and Dowex (50 g/L), at different initial concentrations of VFA. The VFA adsorption process was elucidated using the Freundlich's single- and multi-component models and the Brunauer-Emmet-Teller (BET) model (Eq. 8.7–8.9; ***Table 8.4***). In these models, E is the VFA recovery efficiency (%, expressed for the individual VFA), C_0 is the initial concentration of each VFA (mg/L), C_e is the equilibrium concentration of each VFA in the solution (mg/L), V is the volume of the solution (L), m is the mass of the adsorbent (g), q_e is the amount of adsorbate in the adsorbent under equilibrium conditions (mg/g), Q_0 is the maximum monolayer coverage capacities (mg/g), b is the Langmuir's isotherm constant (L/mg), K_F is the Freundlich's isotherm constant (mg/g) where $(L/g)^n$ is related to the adsorption capacity, n is the adsorption intensity, $C_s = 1/K_L$ (L/mg), q_s is the theoretical isotherm saturation capacity (mg/g) and K_s is the BET equilibrium constant of adsorption for the first layer relating to the energy of surface interaction (L/mg).

8.2.5. Selective adsorption of VFA

Based on the q_e values of Dowex and Amberlite for the individual VFA, a sequential batch adsorption experiment was designed to selectively adsorb the individual VFA. The VFA mixture containing 1g/L of individual VFA (total: 6 g/L) was first equilibrated with 25 g/L of Dowex for 30 min. Thereafter, the liquid-phase was transferred to a 110 mL serum bottle containing 25 g/L of Amberlite and equilibrated for 30 min. For the desorption step, the resins were treated with 100 mM NaOH for 30 min. The liquid samples were analysed for initial and final VFA concentrations after both the adsorption and desorption step.

8.3. Results and discussion

8.3.1. Screening of resins

In this study, 11 different anion-exchange resins and granular activated carbon (GAC) (***Table 8.2***) that were previously examined for the adsorption of anions such as sulfate, selenate (SeO_4^{2-}) (Tan et al., 2018a) and lactate (Pradhan et al., 2017) were examined for VFA adsorption. The percentage adsorption of resins namely, Dowex 1X8-100 (Cl), Dowex 21K XLT, Reillex 425, Dowex marathon A2 (Cl), Dowex ion exchange resin, Amberlite IRA-400 (Cl) and Amberlite IRA-900 (Cl) was < 30%, whereas, the resins such as Amberlite IRA-67 free base, Amberlite IRA-96 free base, Dowex Optipore L-493 and granular activated carbon (GAC) showed adsorption efficiencies > 50% (***Figure 8.1***). GAC showed 50% adsorption for H-ac and > 98% for H-va and the percentage adsorption increased with the aliphatic chain length (ACL) of the VFA. Amberlite IRA-67 (resin no. 7) showed an adsorption > 95% for all VFA, while Amberlite IRA-96 showed adsorption in the range of 75-90%. Both Amberlite IRA-67 and IRA-96 are weakly basic (type 2) resins with polyamine functional group (***Table 8.3***).

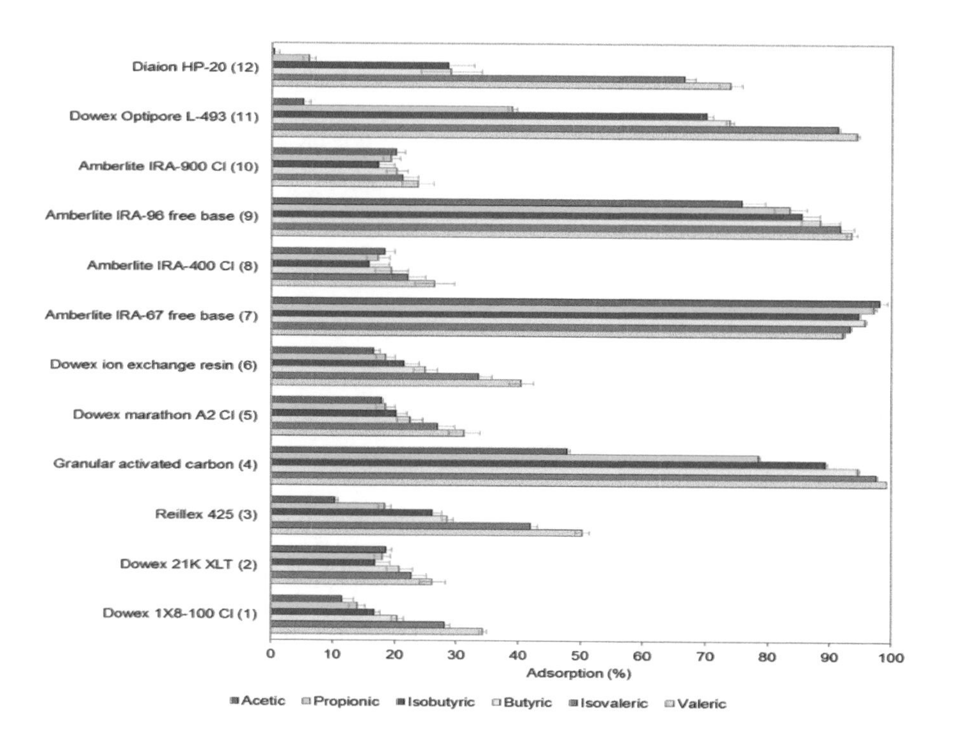

Figure 8.1: *Screening of various commercial resins and GAC for VFA adsorption.*

Among the wide range of adsorbents recommended in the literature for the VFA adsorption (Anasthas and Gaikar, 2001; Rebecchi et al., 2016; Reyhanitash et al., 2017; Tugtas, 2011; Yousuf et al., 2016), type 2 (weakly basic) anion exchange resins are usually preferred because the tertiary amine functional group can adsorb carboxylic acids as charge-neutral units to maintain the charge neutrality (Reyhanitash et al., 2017). Dowex showed $< 20\%$ adsorption for H-ac and up to 90% adsorption for H-va, and the adsorption efficiency increased with the ACL. Based on the resin screening experiment, Amberlite showed the highest adsorption for all the VFA and the VFA adsorption by Dowex was based on the ACL. Hence, Amberlite and Dowex were selected for further experiments on the adsorption of single- and multi-component VFA.

Table 8.2: *Description of anion-exchange resins screened for VFA adsorption.*

	Adsorbate	Manufacturer	Matrix and active site	Type	Active group
1	Dowex 1X8-100 (Cl)	Sigma-Aldrich	styrene-divinylbenzene	type 1, strongly basic	trimethylbenzyl ammonium,
2	Dowex 21K XLT	Supelco	styrene-divinylbenzene	type 1, strongly basic	trimethylbenzyl ammonium
3	Reillex 425	Aldrich	Poly(4-inylpyridine), 25% cross-linked with divinylbenzene		pyridine
4	Granular activated carbon (GAC)				
5	Dowex marathon A2 (Cl)	Sigma-Aldrich	styrene-divinylbenzene	type 2, weakly basic	dimethyl-2-hydroxyethylbenzyl ammonium,
6	Dowex ion exchange resin		standard anion exchange resin		
7	Amberlite IRA-67 free base	Aldrich	styrene-divinylbenzene	type 2, weakly basic	polyamine,
8	Amberlite IRA-400 (Cl)	Aldrich	styrene-divinylbenzene	type 1, strongly basic	trialkylbenzyl ammonium,
9	Amberlite IRA-96 free base	Sigma-Aldrich	styrene-divinylbenzene	type 2, weakly basic	polyamine,
10	Amberlite IRA-900 (Cl)	Sigma-Aldrich	styrene-divinylbenzene copolymer	type 1 strongly basic	trialkylbenzyl ammonium,
11	Dowex Optipore L-493	Sigma-Aldrich	polystyrene- divinylbenzene (PS-DVB)		none
12	Diaion HP-20	Supelco	styrene-divinylbenzene		none

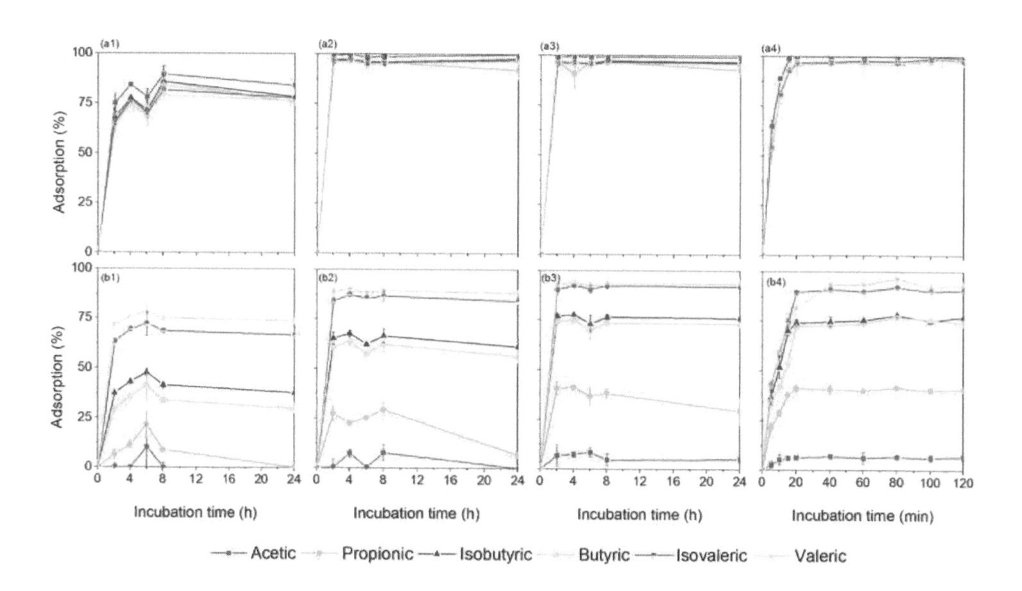

Figure 8.2: *Effect of contact time on the adsorption of VFA by different resins: (a) Amberlite and (b) Dowex at resin concentrations of (1) 25, (2) 50, (3) 75 g/L, respectively. (a4) and (b4) are the profiles of VFA adsorption (%) for Amberlite and Dowex, respectively, at an incubation time of 120 min.*

Table 8.3: *Physical properties of the resins selected for the VFA adsorption studies.*

Ion exchange resin	**Amberlite® IRA-67 free base**	**Dowex Optipore® L-493**
Particle size [mm]	0.50-0.75	0.2-0.3
BET surface area [m^2/g]	8.4	1048
Pore volume [mL/g]	0.007	0.72-0.79
Average pore diameter [nm]	3.3-3.35	2.75-3.0
Bulk density [g/L]	700	680
Total exchange capacity [eq/L]	$\geq$ 1.60	0.6

8.3.2. Effect of resin concentration

In order to study the effect of resin concentration and contact time on VFA adsorption, an aqueous mixture containing 1 g/L of individual VFA (total: 6 g/L) was incubated separately with Amberlite and Dowex for 24 h. ***Figure 8.2(a1–a3)*** shows the percentage adsorption of Amberlite and it is evident that the adsorption of all the 6 VFA in the mixture occurred to the same extent. About 70–80% adsorption of all the VFA occurred for Amberlite at a resin concentration of 25 g/L and at a resin concentration of 50 g/L the adsorption was > 95%.

In the case of Dowex (***Figure 8.2b1–b3***), the percentage adsorption increased with an increase in the resin concentration and the ACL. At a resin concentration of 75 g/L, the percentage adsorption for H-va and H-iv was > 90%, ~ 75% for H-bu and H-ib, ~ 40% for H-pr and < 10% for H-ac, respectively. However, it is noteworthy to mention that for both Amberlite and Dowex the percentage adsorption did not vary much (< 3%) between the straight chain VFA compounds and their isomeric counterparts. Since the effect of the resin concentration on the percentage adsorption also depends on the initial VFA concentration in the liquid-phase, isotherm studies were carried out to determine the adsorption capacities of Amberlite and Dowex.

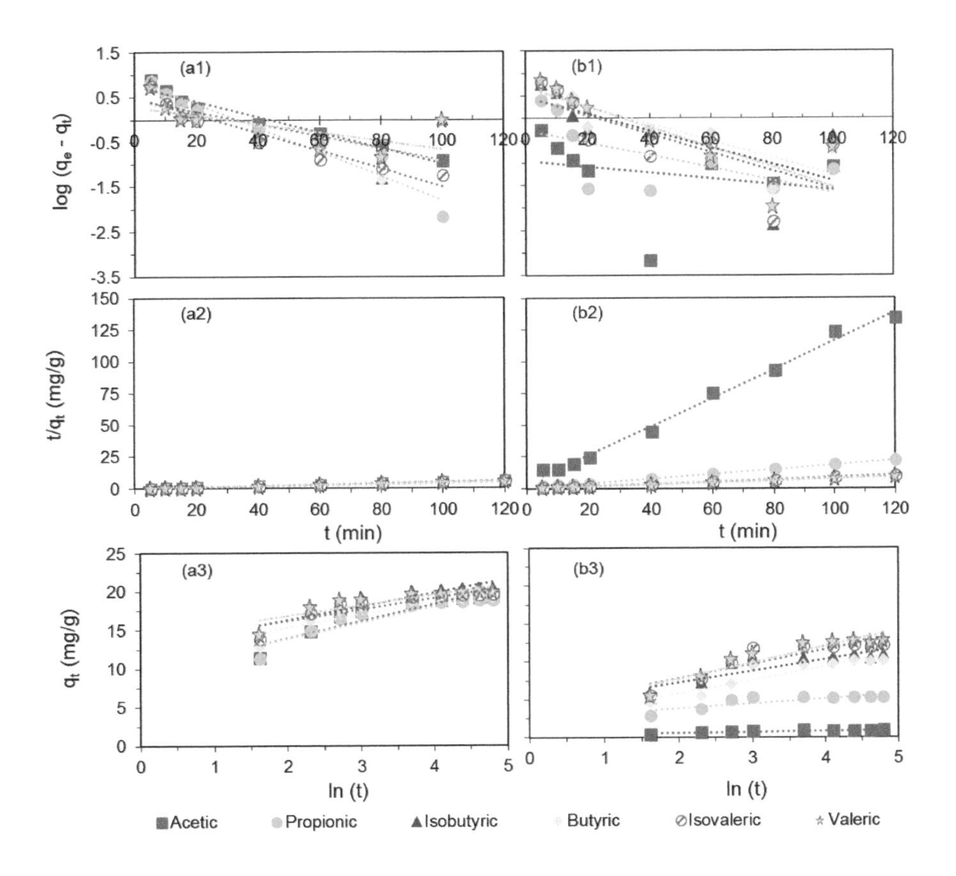

Figure 8.3: *VFA adsorption kinetics data for: (a) Amberlite and (b) Dowex, respectively, fitted to: (1) pseudo-first order, (2) pseudo-second order, and (3) Elovich kinetic model at an initial concentration of 6 g/L of total VFA (1 g/L each) and a resin concentration of 50 g/L of Amberlite and 75 g/L of Dowex. Data points represent the experimental data, while the dotted lines are the model fits.*

Table 8.4: *List of differential and linear form of adsorption kinetic and isotherm model equations*

	Differential form	Linear form	Plot	
Pseudo-first order kinetic model	$\frac{dq_t}{dt} = K_1(q_e - q_t)$	$\log(q_e - q_t) = \log q_e - (\frac{K_1}{2.303})t$	$\log(q_e - q_t)$ vs. t	(Eq. 8.3)
Elovich kinetic model	$\frac{dq_t}{dt} = \alpha e^{-\beta q_t}$	$q_t = \beta \ln \alpha\beta + \ln t$	q_t vs. ln t	(Eq. 8.4)
Pseudo-second order kinetic model	$\frac{dq_t}{dt} = K_2(q_e - q_t)^2$	$\frac{t}{q_t} = \frac{1}{K_2 q_e^2} + \frac{1}{q_e}t = \frac{1}{h_0} + \frac{1}{q_e}t$	$\frac{t}{q_t}$ vs. t	(Eq. 8.5)
Intra-particle diffusion model		$q_t = K_{id}t^{0.5} + C_{id}$	q_t vs. $t^{0.5}$	(Eq. 8.6)
Langmuir single component isotherm model	$q_e = \frac{Q_o b C_e}{1 + bC_e}$	$\frac{C_e}{q_e} = \frac{1}{bQ_o} + \frac{C_e}{Q_o}$	$\frac{C_e}{q_e}$ vs. C_e	(Eq. 8.7)
Langmuir multi-component isotherm model		$q_{ei} = \frac{Q_{oi} b_i C_{ei}}{1 + \sum b_i C_{ei}}$		(Eq. 8.8)
BET (multilayer) isotherm model	$q_e = q_s \frac{K_S C_e}{(C_s - C_e)[1 + (K_S - 1)\frac{C_e}{C_s}]}$	$\frac{C_e}{q_e(C_s - C_e)} = \frac{1}{q_s K_S} + \frac{(K_S - 1)}{q_s K_S} \times \frac{C_e}{C_s}$	$\frac{C_e}{q_e(C_s - C_e)}$ vs. $\frac{C_e}{C_s}$	(Eq. 8.9)

8.3.3. Effect of contact time

For the batches incubated for 24 h, the adsorption equilibrium was achieved and the adsorption was nearly complete within 2 h of incubation at all the three resin concentrations (***Figure 8.2a4 and 8.2b4***). Incubation period of 2 h (Rebecchi et al., 2016) and 1 h (Reyhanitash et al., 2017) is usually recommended to achieve a solid-liquid equilibrium for VFA adsorption. To examine the actual time required to attain equilibrium, batch studies with 50 g/L of Amberlite and 75 g/L of Dowex was performed with an incubation time of 2 h and samples were collected at intervals of 5, 10, 15, 20, 40, 60, 80, 100 and 120 min (***Figure 8.2.a4 and 2.b4***). The percentage adsorption increased exponentially with incubation time for both Amberlite and Dowex, and ~ 50% of the total VFA adsorption occurred within 5 min of incubation and the equilibrium adsorption was achieved within 20 min of incubation. Based on these results, an equilibrium time of 30 min was selected for the subsequent experiments.

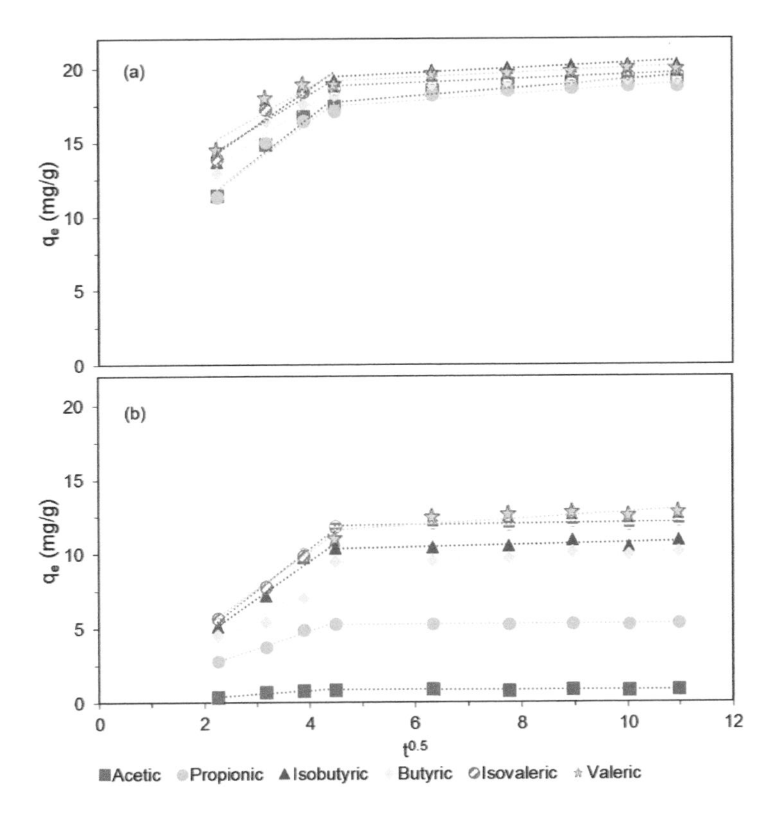

Figure 8.4: *Experimental data fitted to the intraparticle diffusion model for VFA adsorption by: (a) Amberlite and (b) Dowex.*

Table 8.5: *Pseudo-second order model parameters for Amberlite and Dowex.*

	Amberlite						Dowex					
Multi-component VFA	Acetic	Propionic	Isobutyric	Butyric	Isovaleric	Valeric	Acetic	Propionic	Isobutyric	Butyric	Isovaleric	Valeric
q_e [mg/g]	19.82	19.32	20.74	19.53	19.87	20.32	0.89	5.49	11.18	10.75	12.64	13.44
h_0 [mg/g.min]	6.38	6.96	11.33	10.67	13.43	14.14	0.23	2.03	3.01	1.66	3.07	2.65
K_2 [g/mg.min]	0.016	0.018	0.026	0.028	0.034	0.034	0.285	0.067	0.024	0.021	0.019	0.015
R^2	0.99	0.99	0.99	0.99	0.99	0.99	0.99	0.99	0.99	0.996	0.99	0.99

Note: Experiments were performed at room temperature (22 ± 2 °C) with 1 g/L each of the VFA

Table 8.6: *Intra-particle diffusion model to describe the mechanism of VFA (1 g/L each) adsorption by Amberlite and Dowex.*

	Amberlite						Dowex					
	Acetic	Propionic	Isobutyric	Butyric	Isovaleric	Valeric	Acetic	Propionic	Isobutyric	Butyric	Isovaleric	Valeric
K_{id} [mg/g.min$^{-0.5}$]	2.74	2.62	2.46	2.32	2.15	2.01	0.22	1.18	2.48	2.15	2.77	2.52
C_{id} [g/mg]	5.70	5.97	8.80	8.29	9.66	10.73	0.20	1.95	3.08	1.85	2.97	2.89
R^2	0.99	0.97	0.95	0.93	0.95	0.96	0.91	0.98	0.72	0.91	0.99	0.98

8.3.4. Kinetics of VFA adsorption

In order to interpret the nature of adsorption kinetics and its mechanism, the experimental data was fitted to the pseudo-first-order (Eq. 8.3), pseudo-second-order (Eq. 8.5), Elovich (Eq. 8.4) and intraparticle diffusion models. The model fitted profiles for Amberlite and Dowex are shown in ***Figures 8.3a1–a3 and 8.3b1–b3***, respectively. The results showed that the pseudo-second-order kinetics described the VFA adsorption data best with the highest R^2 values (***Table 8.5***), implying that the adsorption of VFA by the resins occurred through chemisorption and the sharing or exchange of electrons between the adsorbate and the adsorbent was the rate-limiting step during the adsorption process (Sheha and El-Shazly, 2010). From the linearized plot of t/q_t vs. t (***Figure. 8.3a2 and 8.3b2***), the pseudo-second-order constants namely q_e, h_o and K_2 were determined (***Table 8.5***). For Amberlite, the q_e, i.e. the equilibrium adsorption capacity, values of all the VFA were similar and in the range of 19.32–20.7 mg/g, whereas for Dowex, the q_e values varied between 1.31 (H-ac) and 19.98 (H-va) mg/g. The K_2 values (pseudo-second-order constant) for Amberlite followed the order: H-va and H-iv > H-bu and H-ib > H-pr > H-ac. Concerning Dowex, the K_2 value for H-ac was the highest (0.258 g/mg.min) and least for H-va (0.015 g/mg.min). Comparing the h_0 and K_2 of H-ac and H-va, the adsorption rate of H-ac (H-ac showed a much higher K_2 value) was not proportional to the number of unoccupied sites, which is not consistent with the observed low q_e value. Furthermore, high h_o for H-ac suggest a rather fast adsorption process, which should correspond to a high q_e value compared to other VFA. Nevertheless, the observed low q_e value for H-ac when compared to other VFA implies that the exchange of electrons at the surface of the resin was not the rate limiting step for the adsorption of H-ac, rather the mechanism through which the molecules are transported to the charged surface of the resin.

8.3.5. Diffusion mechanism for VFA adsorption

Synthetic ion-exchange resins are usually macroporous polymeric material with a large surface area (e.g. surface area of Dowex: ~ 1100 m^2/g). The functional groups of these resins are located in the pores as well as on the surface. In completely stirred batch adsorption experiments, the mass transfer characteristics can be related to the diffusion coefficient by fitting the experimental adsorption rate data to diffusion based models. In such systems, the intraparticle diffusion model can help in understanding the diffusion mechanism of the adsorbate from the bulk liquid-phase to the resin surface (Ahmaruzzaman, 2008; Sheha and El-Shazly, 2010). From an engineering viewpoint, the sorption process is controlled by diffusion if its rate depends on the rate at which the adsorbate diffuse towards the charged surface site on the resin. The linear form of the intraparticle diffusion model can be visualised by plotting q_e vs. $t^{0.5}$. If the plot is a straight line, the adsorption is controlled solely by intraparticle diffusion. On the other hand, if the data exhibits multi-linear plots, two or more steps could influence the adsorption process (Sheha and El-Shazly, 2010; Sun et al., 2015). From ***Figures 8.4a and 8.4b***, it is evident that the data points correspond to two straight lines, with the first line representing macropore diffusion and the second line representing micropore diffusion (Sheha and El-Shazly, 2010; Tan et al., 2018a). The intraparticle diffusion rate constant K_{id} (mg/g.min$^{-0.5}$) and C_{id} (mg/g), which gives an approximation of the boundary layer thickness, were calculated using the intra-particle diffusion model (Eq. 8.6). Apparently, both K_{id} and C_{id} increased with the ACL for both Amberlite and Dowex (***Table 8.6***). At the tested initial VFA concentrations, the magnitude of the difference in K_{id} and C_{id} values between the shortest (H-ac) and the longest (H-va) VFA were much higher for Dowex, while the values were comparable for Amberlite.

8.3.6. Desorption and VFA recovery

For the successful application of VFA adsorption at an industrial scale, recovery of the adsorbed VFA and regeneration of the resins through desorption is crucial. In this study, desorption of ~ 1000 mg/L of individual VFA using different molar strengths of NaOH (45, 60, 70, 75, 80, 100 mM) was tested. ***Figure 8.5a–f*** shows the profiles of percentage desorption of VFA from Amberlite. It is evident that the desorption efficiency increased with an increase in the molar strength of NaOH, simply due to the increased availability of OH^- ions to displace the VFA molecules from the active site. The NaOH strength was selected based on the cumulative molar strength of the VFA mixture (~ 73 mM). Since a desorption efficiency > 98% for all the VFA was achieved using 100 mM of NaOH, it is evident that the VFA desorption through ionic displacement was driven by the concentration gradient of the adsorbing ions at equilibrium. Desorption was almost instantaneous with most of it occurring within the first 5 min of incubation and desorption was almost complete within 15–20 min of incubation. Based on these results, an incubation time of 30 min was selected for future experiments involving desorption.

The desorption techniques typically reported in the literature involves washing with solvents, e.g. water (Guerra et al., 2014), methanol (Cao et al., 2002), ethanol and propanol (Da Silva and Miranda, 2013), mixture of ethanol and NaOH (Rebecchi et al., 2016), acidic or alkali solution (Guerra et al., 2014). For the recovery of high purity compounds, the desorbate should be subjected to a post-treatment step (e.g. distillation for organic acids). Pressure or temperature swing desorption along with nitrogen stripping can also be used for the recovery of VFA from the resins (Da Silva and Miranda, 2013). For future studies, additional experiments should be conducted to examine the effect of multiple adsorption-desorption cycles on the VFA recovery and the resin stability, and the application of desorption models to elucidate the kinetics and mechanism of the desorption process.

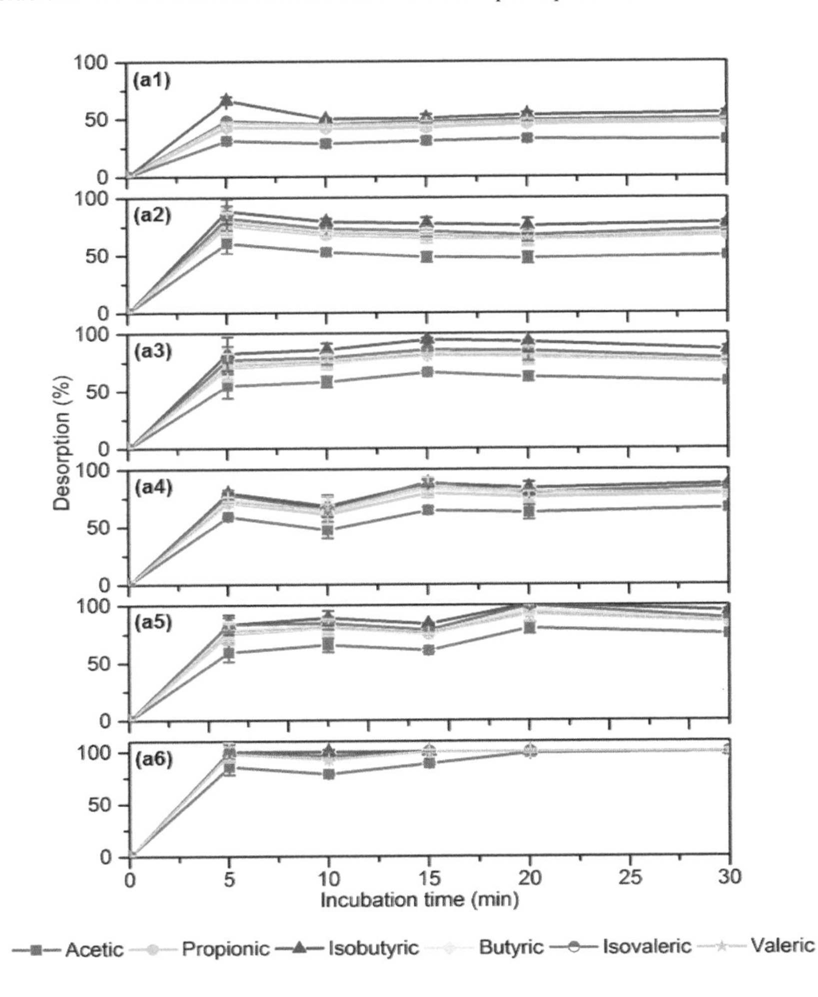

Figure 8.5: *Desorption of VFA from Amberlite using: (a) 45, (b) 60, (c) 70, (d) 75, (e) 80, and (f) 100 mM NaOH.*

8.3.7. Adsorption isotherm

Langmuir's isotherm for single- and multi-component system

The equilibrium VFA concentration and the respective adsorption capacity values (q_e) at initial concentrations ranging between 0.1 and 10 g/L in single-component batch systems were fitted to the Langmuir's and Freundlich's single-component isotherms. The primary assumptions for the Langmuir's adsorption isotherm model are: (i) the adsorption occurs as a monolayer, (ii) the adsorption sites are finite, localised, identical and equivalent, and (iii) there is no lateral interaction or steric hindrance between the adsorbed molecules, nor between the adjacent sites (Foo and Hameed, 2010; Guerra et al., 2014). The Freundlich's isotherm describes the non-ideal, reversible adsorption, and it could be applied for multi-layer adsorption over heterogeneous surfaces, wherein the amount adsorbed is the sum of adsorption that occurs at all the sites. The stronger sites are occupied first until the adsorption energy is exponentially decreased, and thereafter adsorption continues at a rather slower rate at the weaker sites [25].

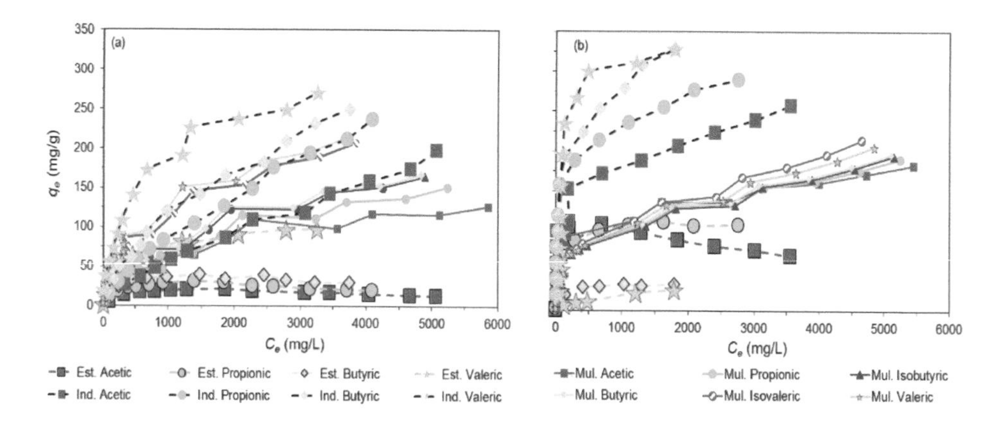

***Figure 8.6**: Adsorption capacity (q_e) vs equilibrium concentration (C_e) plot for: (a) Amberlite and (b) Dowex. Comparison between q_e in single component batch system is denoted with prefix 'Ind.', estimated q_e for a multi-component system is denoted with the prefix 'Est.', and the experiment based q_e in a real multi-component batch system is denoted with the prefix 'Mul.'.*

Table 8.7: *Langmuir single-component and multi-component, and BET isotherm model parameters for Amberlite and Dowex.*

	Langmuir model for single-component system											
	Single component system						Estimated values for multi component system					
	Amberlite			Dowex			Amberlite			Dowex		
	q_m [mg/g]	$B \times 10^{-3}$ [L/mg]	R^2 -	q_m [mg/g]	$B \times 10^{-3}$ [L/mg]	R^2 -	q_m [mg/g]	$B \times 10^{-2}$ [L/mg]	R^2 -	q_m [mg/g]	$B \times 10^{-3}$ [L/mg]	R^2 -
Acetic acid	246	7.52	0.98	218	0.4	0.99	75	3.05	0.99	13.4	3.3	0.97
Propionic acid	282	6.31	0.99	227	0.52	0.99	110.2	2.01	1.0	21.0	5.8	0.98
Isobutyric acid	-	-	-	-	-	-	-	-	-	-	-	-
Butyric acid	336	7.17	0.98	247	0.98	0.97	33.1	4.34	0.99	31.7	21.4	0.98
Isovaleric acid	-	-	-	-	-	-	-	-	-	-	-	-
Valeric acid	360	6.24	0.97	263	2.76	0.98	25.2	1.02	0.94	102.3	4.24	0.99

	Multi-component system													
	Langmuir model						BET model							
	Amberlite			Dowex			Amberlite				Dowex			
	q_m [mg/g]	$B \times 10^{-3}$ [L/mg]	R^2 -	q_m [mg/g]	$B \times 10^{-3}$ [L/mg]	R^2 -	q_s [mg/g]	$K_S \times 10^{-3}$ [L/mg]	$K_L \times 10^{-5}$ [L/mg]	R^2	q_s [mg/g]	$K_S \times 10^{-3}$ [L/mg]	$K_L \times 10^{-5}$ [L/mg]	R^2
Acetic acid	169.4	9.38	0.98	124.1	0.99	0.93	115.3	38.11	7.00	0.99	76.0	1.75	7.2	0.98
Propionic acid	175.4	4.36	0.96	161.0	0.91	0.89	106.2	12.2	8.77	0.99	87.6	1.70	8.62	0.98
Isobutyric acid	178.1	3.91	0.95	172.1	1.24	0.91	104.1	10.9	9.40	0.99	93.4	2.73	9.52	0.98
Butyric acid	179.6	4.52	0.96	168.2	1.62	0.92	104.8	14.38	9.52	0.99	92.9	3.92	9.75	0.98
Isovaleric acid	195.3	4.27	0.95	199.0	3.19	0.93	111.3	12.68	10.8	0.99	114.5	8.58	12.2	0.99
Valeric acid	189.4	3.77	0.95	209.1	3.08	0.95	107.6	10.61	10.3	0.99	122.3	7.74	12.27	0.99

The adsorption data of VFA on Dowex showed a good fit for both Freundlich and Langmuir models, whereas the adsorption data on Amberlite fitted the Freundlich model with a low R^2 value (< 85%). This observation could be attributed to the fact that VFA adsorption occurred in multi-layers, especially on Amberlite and a single layer model such as Freundlich was not able to explain the higher adsorption capacity (e.g. 169.4 and 124.1 mg/g for H-Ac with Amberlite and Dowex, respectively) than the Freundlich model prediction (e.g. 75 and 25.2 mg/g for H-Ac with Amberlite and Dowex, respectively). To compare the adsorption between Amberlite and Dowex in single- and multi-component systems, the Langmuir's isotherm model was chosen for further investigation and the profiles (data not shown) of the Langmuir's isotherm for Dowex and Amberlite were obtained using the linear form of Eq. 8.7; ***Table 8.4***. The model parameters namely the maximum adsorption capacity, q_e (mg/g) and b (L/mg), which represents a criteria of the adsorbent-adsorbate affinity for single-component systems were determined (***Table 8.7***).

In the single-component system, the q_m for both Amberlite and Dowex increased with the ACL of the VFA with a value of 246 and 218 mg/g for H-ac and 380 and 263 mg/g for H-va, respectively. The b values for Amberlite were in a narrow range of 6.24-7.52 × 10^{-3} L/mg, while for Dowex, the b values were in the range of 0.4 (H-ac) – 2.76 (H-va) × 10^{-3} L/mg. These results were in accordance with Traube's rule which relates the adsorption process to the surface tension. Accordingly, in homologous series, VFA cause a decrease in the surface tension to a larger degree as the length of the carbon chain increases. Besides, organic acids tend to adsorb with the carbon chain parallel to the surface of the adsorbent and each $-CH_2-$ group contributes the same adsorption energy to the molecule [30]. Although the Traube's rule is more relevant to physical adsorption on adsorbents such as activated carbon, for the chemisorption of carboxylic acids on ion exchange resins, the additional increase in the adsorption energy with increasing carbon chain length increases the adsorption capacity of the resins (Bansal and Goyal, n.d.; Da Silva and Miranda, 2013).

Based on the q_m and b values obtained from the single-component system, the q_e values for the multi-component system were estimated using an extended, predictive Langmuir model (Eq. 8.7; ***Table 8.4***). This model assumes a homogeneous surface with respect to the energy of adsorption, absence of interaction between the adsorbed species and that all the active sites are equally available for the adsorbate [17,32]. In case of Amberlite, the estimated q_m value was in the order of H-pr > H-ac > H-bu > H-va, whereas it increased with ACL for Dowex. However, neither the magnitude nor the trend of the estimated q_m and b values corresponded to the q_m and b values calculated from the experimental data (***Table 8.7***).

Furthermore, in the multi-component batch systems for Amberlite, the q_m values for all VFA varied in a narrow range (169.4 and 195.3 mg/g), compared to the single-component system (246–360 mg/g). The b values were in the range of 3.7–9.38 $\times$ 10^{-3} L/mg for Amberlite and 0.99–3.19 $\times$ 10^{-3} L/mg for Dowex, wherein H-va and H-ac had the least and the highest value, respectively. The Langmuir constants (q_m and b) were ~ 2 to 8 times higher than the estimated values indicating that the adsorption occurred in multi-layers. A comparison of the q_e for individual VFA in single-component system, the estimated q_e for the multi-component and the actual q_e values for the multi-component system are shown in ***Figures 8.6a and 8.6b***. Based on the fact that the q_m value was ~ 2–8 times higher than the estimated value and a lower R^2 value for Dowex, a multi-layer adsorption model was considered to describe the adsorption process.

BET isotherm

The BET model (Eq. 8.9; ***Table 8.4***) was applied to the multi-component batch adsorption system data using the experimentally determined q_e and c_e values. The BET model plots are shown in ***Figure 8.7a and 8.7b***. By comparing the R^2 values for the Langmuir and the BET model, it is evident that the BET model was more suitable to describe the adsorption process in batch system. The BET constants were calculated based on the method proposed by Ebadi et al. [33]. In the BET equation for gas-phase adsorption, the equilibrium constant of adsorption for the outer layer (K_L) was fixed by expressing the adsorbate concentration in the form of a relative concentration, i.e. by equating $1/K_L$ to the saturation pressure, P_S. By doing so, the number of parameters was reduced to two, i.e. q_m and K_S, and the equation was linearized. In the case of liquid-phase adsorption, it was reported that the adsorption capacity (q) does not tend towards infinity when the adsorbent is saturated [33]. Besides, adsorption will also not occur at the saturation concentration (C_s) if the surface of the adsorbent has no affinity towards the adsorbate. On the contrary, if the affinity is very high, the q values could be high despite the low adsorbent dose (Ebadi et al., 2009). Thus, in the present study, BET isotherm equation (Eq. 8.9), C_s does not represent the actual saturation concentration, but it is an adjustable parameter which is the inverse of K_L (equilibrium or affinity constant for upper layers, L/mg). The equilibrium constant of adsorption for the first layer K_S (L/mg) is given by $K_S = C_{BET}/C_S$.

The maximum adsorption capacities for the VFA calculated using the BET model (q_s for H-ac = 115.3 mg/g) was lower than the value calculated using the Langmuir model (q_m for H-ac =

169.4 mg/g). This observation could be verified by considering the physical significance of K_S and K_L. The constant b in the Langmuir isotherm, which establishes the affinity between the absorbent and the adsorbate molecules singularly represents the affinity of interaction because the Langmuir's model is intended to represent a monolayer adsorption. The layer of adsorbent molecules around the adsorbate is taken as a single uniform layer, leading to an overestimation of b compared to the BET model. The latter model assumes that for each adsorption site, an arbitrary number of molecules may be accommodated, and hence, adsorption is a multilayer process and the layers need not necessarily be uniform. Comparing the K_S and K_L (equilibrium constant for the first layer and the outer layers, respectively), it is evident that the K_L is ~ 100 times less than the K_S. This implies that the affinity greatly reduced in the outer layers, leading to lower maximum adsorption capacity values. In the case of Amberlite, the q_m values for different VFA were nearly similar and in the range of 104.1–111.3 mg/g, whereas for Dowex, the q_m increased with the ACL, e.g. 76 g/g for H-ac and 122.3 g/g for H-va. A similar trend was observed for the Ks and K_L values, where the values were within a similar range for Amberlite (except for H-ac), where as they increased with the increasing ACL for Dowex.

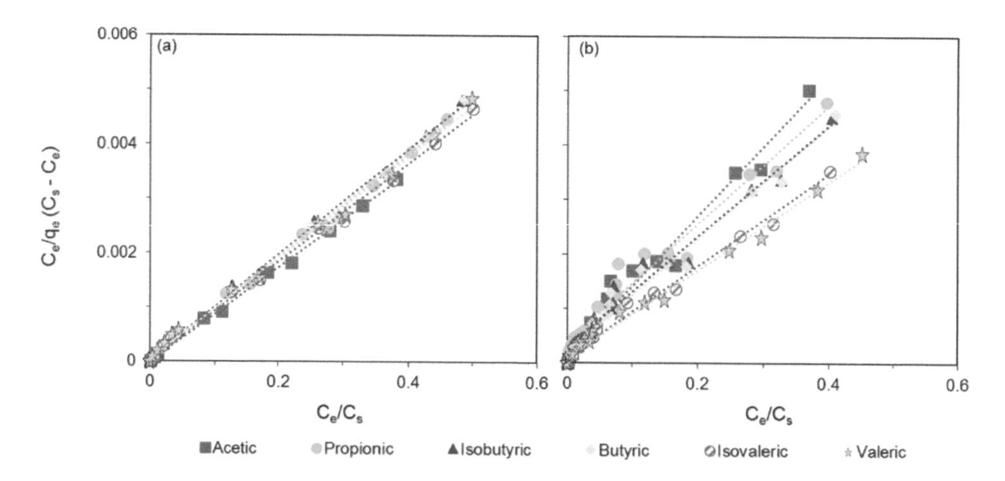

Figure 8.7: *Equilibrium adsorption isotherm data fitted to the BET multi-component model for: (a) Amberlite and (b) Dowex.*

8.3.8. Sequential batch adsorption for separation of individual VFA

In industrial situations, the separation of individual VFA from each other and/or from other organic acids present in the fermentation broth still remains a major challenge (Eregowda et al., 2018). Based on the q_m values for VFA adsorption on Amberlite and Dowex, a sequential adsorption process was designed. ***Figure 8.8*** shows the percentage adsorption and desorption on Dowex and Amberlite, respectively, for a mixture containing 1000 mg/L of each VFA in a sequential batch adsorption process. It can be clearly seen that ~ 80% H-va and H-iv were recovered from Dowex, while > 85% H-ac and ~ 75% of H-pr was recovered from Amberlite. In the case of H-bu and H-ib, 55–60% was adsorbed by Dowex and the remaining was recovered by adsorption onto Amberlite. The solid brown and yellow colouration seen in ***Figure 8.8*** represents the fraction that was lost during the desorption step.

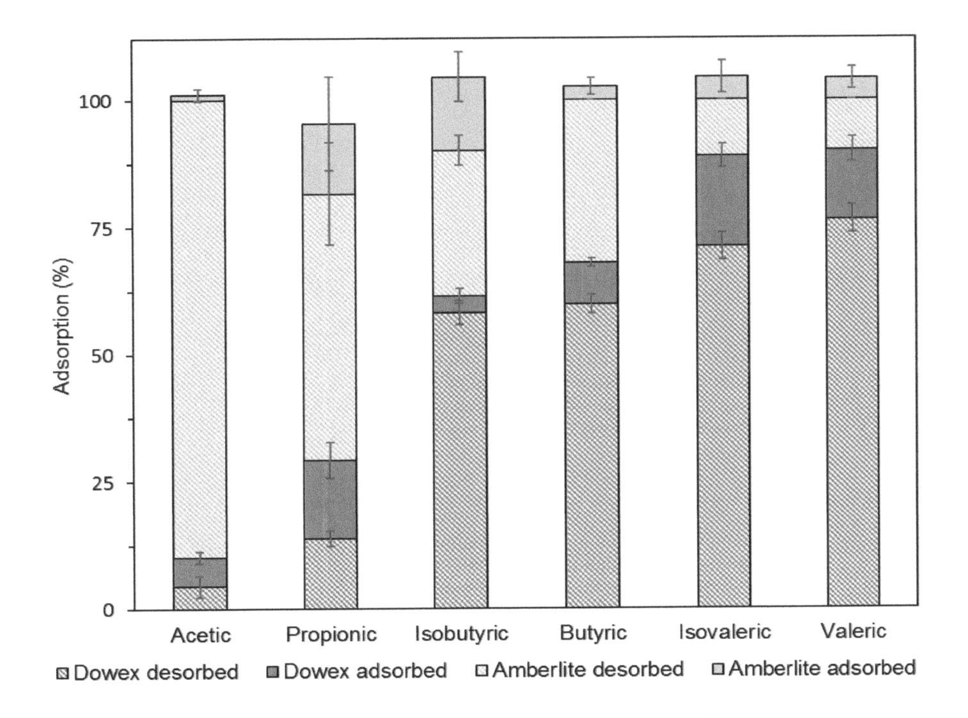

Figure 8.8: *Separation of acetic acid due to selective adsorption of VFA on Amberlite and Dowex.*

The selective separation of H-ac from the VFA mixture was driven by the differential affinity of individual VFA, especially H-ac towards Dowex. As demonstrated, under the tested conditions, Amberlite could almost completely adsorb all the tested VFA and through a combination of resins (Dowex in this case), different adsorption capacities for each VFA as compared to the q_e for individual resin were achieved. The VFA fraction that was not adsorbed by Dowex adsorbed onto Amberlite and thus, nearly complete recovery of the VFA mixture was possible. ***Figure 8.2 (b1-b3)*** shows a clear trend of an increase in percentage adsorption of H-ac and H-pr with an increase in the resin increasing concentration for Dowex under equilibrium conditions. For scale-up purposes, the VFA to resin ratio could also be considered in order to achieve selective adsorption of the VFA. Based on the results obtained from this batch study, the sequential adsorption process could be extended to continuous column studies along with the optimization of the reactors hydrodynamic and operational parameters. By using a combination of resins, the sequential adsorption process could be extended for the adsorption of ionic compounds like sulfate, selenate and other organic acids. The challenge of interference of similarly charged ions (for example, sulfate and chloride ions interfering in the adsorption of VFA) could be overcome by selective separation through sequential batch adsorption using two or more resins, provided with appropriate choice of resin. As reported in the literature, the Type 2 weak anion exchange resins, e.g. Amberlite IRA-67, are good candidates for the adsorption of anions such as selenate, sulphate (Tan et al., 2018a) (Coma et al., 2016), lactic acid (Bayazit et al., 2011; Yousuf et al., 2016), H-ac and H-bu (Yousuf et al., 2016). Polystyrene- divinylbenzene resins, e.g. Dowex Optiopore L-493, which has high preference for H-ac as demonstrated in this study, and Lewatit VP OC 1064 MD PH, which is selective to adsorb H-Bu (Reyhanitash et al., 2017), can also be considered for the selective separation of individual VFA or other anions.

8.4. Conclusions

This study explored the adsorption of a mixture of VFA containing H-ac, H-pr, H-ib, H-bu, H-iv and H-va on to anion exchange resins. Anion exchange resins suited for VFA adsorption were screened and the kinetics and the adsorption capacity of two selected resins (namely Amberlite IRA and Dowex Optipore L-493) were compared. Based on the adsorption capacities of the resins for different VFA, selective separation of individual VFA was possible using a sequential batch adsorption step involving two resins. From a practical viewpoint, for *in-situ* product recovery, the sequential batch adsorption process could also be further extended for the recovery and purification of prospective second-generation biofuels.

Acknowledgements

The authors thank ass. prof. Marika E. Kokko (TUT, Finland) for her suggestions and comments during the planning of the adsorption experiments. This work was supported by the Marie Skłodowska-Curie European Joint Doctorate (EJD) in Advanced Biological Waste-to-Energy Technologies (ABWET) funded by Horizon 2020 under the grant agreement no. 643071.

References

Ahmaruzzaman, M. (2008) Adsorption of phenolic compounds on low-cost adsorbents: A review, Adv. Colloid Interface Sci., 143: 48–67.

Anasthas, H.M.; Gaikar, V.G. (2001) Adsorption of acetic acid on ion-exchange resins in non-aqueous conditions, React. Funct. Polymers., 47: 23–35.

Baig, K.S.; Doan, H.D.; Wu, J. (2009) Multicomponent isotherms for biosorption of Ni^{2+} and Zn^{2+}, Desalination., 249: 429–439.

Bansal, R.C.; Goyal, M. (2005) Activated Carbon Adsorption, CRC Press, Boca Raton, USA, pp: 520.

Baumann, I.; Westermann, P. (2016) Microbial production of short chain fatty acids from lignocellulosic biomass: current processes and market, Biomed Res. Int. ID 8469357.

Bayazit, Ş.S.; Inci, I.; Uslu, H. (2011) Adsorption of lactic acid from model fermentation broth onto activated carbon and amberlite IRA-67, J. Chem. Eng. Data., 56: 1751–1754.

Cao, X.; Yun, H.S.; Koo, Y.M. (2002) Recovery of L-(+)-lactic acid by anion exchange resin Amberlite IRA-400, Biochem. Eng. J., 11: 189–196.

Choudhari, S.K.; Cerrone, F.; Woods, T.; Joyce, K.; O'Flaherty, V.; O'Connor, K.; Babu, R. (2015) Pervaporation separation of butyric acid from aqueous and anaerobic digestion (AD) solutions using PEBA based composite membranes, J. Ind. Eng. Chem., 23: 163–170.

Da Silva, A.H.; Miranda, E.A. (2013) Adsorption/desorption of organic acids onto different adsorbents for their recovery from fermentation broths, J. Chem. Eng. Data., 58: 1454–1463.

Djas, M.; Henczka, M. (2018) Reactive extraction of carboxylic acids using organic solvents and supercritical fluids: A review, Sep. Purif. Technol., 201: 106–119.

Ebadi, A.; Soltan Mohammadzadeh, J.S.; Khudiev, A. (2009) What is the correct form of BET isotherm for modelling liquid phase adsorption?, Adsorption., 15: 65–73.

Eregowda, T.; Matanhike, L.; Rene, E.R.; Lens, P.N.L. (2018) Performance of a biotrickling filter for anaerobic utilization of gas-phase methanol coupled to thiosulphate reduction and resource recovery through volatile fatty acids production, Bioresour. Technol., 263: 591–600.

Foo, K.Y.; Hameed, B.H.; (2010) Insights into the modeling of adsorption isotherm systems. Chem. Eng. J., 156: 2-10.

Guerra, D.J.L.; Mello, I.; Freitas, L.R.; Resende, R.; Silva, R.A.R. (2014) Equilibrium, thermodynamic, and kinetic of Cr(VI) adsorption using a modified and unmodified bentonite clay, Int. J. Min. Sci. Technol., 24: 525–535.

Jones R.J.; Massanet-Nicolau, J.; Mulder, M.J.J.; Premier, G.; Dinsdale, R.; Guwy, A.; (2017) Increased biohydrogen yields, volatile fatty acid production and substrate utilisation rates via the electrodialysis of a continually fed sucrose fermenter, Bioresour. Technol., 229: 46–52.

Jones, R.J.; Massanet-Nicolau, J.; Guwy, A.; Premier, G.C.; Dinsdale, R.M.; Reilly, M. (2015) Removal and recovery of inhibitory volatile fatty acids from mixed acid fermentations by conventional electrodialysis, Bioresour. Technol., 189: 279–284.

Longo, S.; Katsou, E.; Malamis, S.; Frison, N.; Renzi, D.; Fatone, F. (2014) Recovery of volatile fatty acids from fermentation of sewage sludge in municipal wastewater treatment plants, Bioresour. Technol., 175: 436–444.

López-garzón, C.S.; Straathof, A.J.J. (2014) Recovery of carboxylic acids produced by fermentation, Biotechnol. Adv., 32: 873–904.

Martinez, G.A.; Rebecchi, S.; Decorti, D.; Domingos, J.M.B.; Natolino, A.; Del Rio, D.; Bertin, L.; Da Porto, C.; Fava, F. (2015) Towards multi-purpose biorefinery platforms for the valorisation of red grape pomace: production of polyphenols, volatile fatty acids, polyhydroxyalkanoates and biogas, Green Chem., 261–270.

Pan, X.R.; Li, W.W.; Huang, L.; Liu, H.Q.; Wang, Y.K.; Geng, Y.K.; Kwan-Sing Lam, P.; Yu, H.Q. (2018) Recovery of high-concentration volatile fatty acids from wastewater using an acidogenesis-electrodialysis integrated system, Bioresour. Technol., 260: 61–67.

Pradhan, N.; Rene, E.R.; Lens, P.N.L.; Dipasquale, L.; D'Ippolito, G.; Fontana, A.; Panico, A.; Esposito, G. (2017) Adsorption behaviour of lactic acid on granular activated carbon and anionic resins: Thermodynamics, isotherms and kinetic studies, Energies., 10: 1–16.

Rebecchi, S.; Pinelli, D.; Bertin, L.; Zama, F.; Fava, F.; Frascari, D. (2016) Volatile fatty acids recovery from the effluent of an acidogenic digestion process fed with grape pomace by adsorption on ion exchange resins, Chem. Eng. J., 306: 629–639.

Reyhanitash, E.; Kersten, S.R.A.; Schuur, B. (2017) Recovery of volatile fatty acids from fermented wastewater by adsorption, ACS Sustain. Chem. Eng., 5: 9176–9184.

Reyhanitash, E.; Zaalberg, B.; Kersten, S.R.A.; Schuur, B. (2016) Extraction of volatile fatty acids from fermented wastewater, Sep. Purif. Technol., 161: 61–68.

Sheha, R.R.; El-Shazly, E.A. (2010) Kinetics and equilibrium modeling of Se(IV) removal from aqueous solutions using metal oxides, Chem. Eng. J., 160: 63–71.

Sun, W.; Pan, W.; Wang, F.; Xu, N. (2015) Removal of Se(IV) and Se(VI) by MFe2O4 nanoparticles from aqueous solution, Chem. Eng. J., 273: 353–362.

Tan, L.C.; Calix, E.M.; Rene, E.R.; Nancharaiah, Y.V.; van Hullebusch, E.D.; Lens, P.N.L. (2018) Amberlite IRA-900 ion exchange resin for the sorption of selenate and sulfate: Equilibrium, Kinetic, and regeneration studies, J. Environ. Eng., 144:

Tugtas, A.E. (2011) Fermentative organic acid production and separation, Fen Bilim. Derg., 23: 70–78.

Uslu, H.; Bayazit, S. (2010) Adsorption equilibrium data for acetic acid and glycolic acid onto amberlite, J. Chem. Eng. Data., 298: 1295–1299.

Wu, W.L.; Zhou, Q.; Yuan, L.; Deng, X.L.; Long, S.R.; Huang, C.L,; Cui, H.P.; Huo, L.S.; Zheng, J.X. (2016) Equilibrium, kinetic, and thermodynamic Studies for crude structured-lipid deacidification using strong-base anion exchange resin, J. Chem. Eng. Data., 61: 1876–1885.

Yousuf, A.; Bonk, F.; Bastidas-Oyanedel, J.R.; Schmidt, J.E. (2016) Recovery of carboxylic acids produced during dark fermentation of food waste by adsorption on Amberlite IRA-67 and activated carbon, Bioresour. Technol., 217: 137–140.

Zhang, Y.; Angelidaki, I. (2015) Bioelectrochemical recovery of waste-derived volatile fatty acids and production of hydrogen and alkali, Water Res., 81: 188–195.

Zhou, M.; Yan, B.; Wong, J.W.C.; Zhang, Y. (2017) Enhanced volatile fatty acids production from anaerobic fermentation of food waste: A mini-review focusing on acidogenic metabolic pathways, Bioresour. Technol., 248: 68–78.

CHAPTER 9

General discussion, conclusions

9.1. General discussion

The paper industry is the fourth largest primary energy consumer with an estimate of 8 exajoules (EJ) per year and up to 28.8% of this energy is consumed for the black liquor and wood waste management (Bajpai, 2016). Out of the 37.8 million tonnes of pulp produced globally in the year 2017, 27.6 million tonnes pulp (73.1%) were produced through the chemical pulping process (CEPI, 2017), and every tonne of produced pulp contributes to ~ 1.5 tonnes black liquor (Tran and Vakkilainnen, 2012). Although the recovery of pulping chemicals (NaOH and Na_2S) and heat is carried out efficiently through the combustion of organics, huge amounts of combustion gases (CO and CO_2), volatile organic and reduced sulfur compounds (such as methanol, ethanol, acetone, alpha-pinene, formaldehyde, methyl mercaptan, dimethyl, sulfide and dimethyl disulphide) are released to the atmosphere. Methanol is an important component of the atmospheric emissions and up to 36000 tonnes of methanol were emitted from the Kraft mills of the pulp and paper industries in the United States alone in the year 2017 (TRI Program, 2017). Given the enormity of the pulping industry, there is a huge potential for biological treatment and resource recovery from the gaseous and liquid effluents such as non-condensable gases (NCG) and Kraft condensate streams from the chemical pulping process of a pulp and paper industry.

Resource recovery from the waste effluents, e.g. production of waste-derived-volatile fatty acids (VFA), is gaining a lot of attention from researchers due to their application in the production of fodder, disinfectants, tanning agents and as precursors for biopolymers and biofuels (Da Silva and Miranda, 2013; Rebecchi et al., 2016). Waste-derived-VFA are common by-products or intermediates during the anaerobic digestion and fermentation of organics. Considering the fact that the majority of VFA is produced petrochemically, waste-derived-VFA production holds a huge potential and waste streams such as kitchen waste, food waste, organic fraction of the municipal solid waste, manure, lingo-cellulosic biomass, primary and waste activated sludge, effluents from paper mill, olive oil mill, wood mill, palm oil mill, wastewaters from cheese whey and dairy and other food processing industries can be used as bio based raw materials (Baumann and Westermann, 2016; Lee et al., 2014). The literature on the bioreactors and bioprocesses for the treatment and resource recovery from methanol-rich gaseous and liquid effluents is discussed in detail in **Chapter 2**.

In order to tackle emissions of methanol-rich gaseous and liquid effluents, lab scale biosystems such as biotrickling filters (BTF), upflow anaerobic sludge blanket (UASB) reactors and batch systems were tested to produce VFA under different operating conditions (**Chapters 3, 4, 5, 6 and 7**). Anaerobic methylotrophy for methane vs VFA production in the presence of different electron acceptors such a sulfate, thiosulfate and selenate was compared in batch microcosm studies in **Chapter 3**. Concerning waste gases such as methanol and methane, its removal efficiency and VFA production potential was tested in BTFs at different loading rates in the presence of thiosulfate and selenate (**Chapters 4, 5 and 6**). Operating parameters with respect to electron acceptor feeding such as step-feed and continuous-feed (**Chapter 4**) and the effect of trickling liquid feed variations, such as intermittent liquid trickling in 6 and 24 h cycles and operation without pH regulation were evaluated (**Chapter 5**).

Three UASB reactors were operated at 22, 37 and 55 °C to examine and compare the methanol utilization from foul condensate (FC), one of the Kraft condensate streams from chemical pulping mills with high concentrations of methanol and reduced sulfur compounds, for VFA production. Prior to the operation of the reactors with the FC feed, acetogenic biomass was enriched in the three UASB reactors by feeding methanol-rich synthetic wastewater. Batch activity assays were performed using biomass collected from the UASB reactors in order to compare the methanol utilization, methane and VFA production rates before and after feeding the UASB reactor with FC (**Chapter 7**). VFA recovery was examined in **Chapter 8** using two anion-exchange resins namely Amberlite IRA-67 and Dowex optipore L-493. The adsorption kinetics and adsorption isotherms were evaluated. Furthermore, a sequential batch process was tested to achieve separation of acetate from the VFA mixture first using Dowex optipore L-493, which adsorbed isovalerate, valerate, isobutyrate and butyrate completely and propionate partially, followed by the adsorption of acetate and the leftover fraction of propionate on Amberlite IRA-67.

An overview of the results for each research objective along with the future perspectives of these technologies based on the findings from this thesis is illustrated in ***Figure 9.1***. ***Figure 9.2*** is an overview of the application of the results obtained from this thesis for the anaerobic treatment and resource recovery through the production and recovery of waste-derived VFA from the gaseous and liquid effluents of a Kraft pulping process.

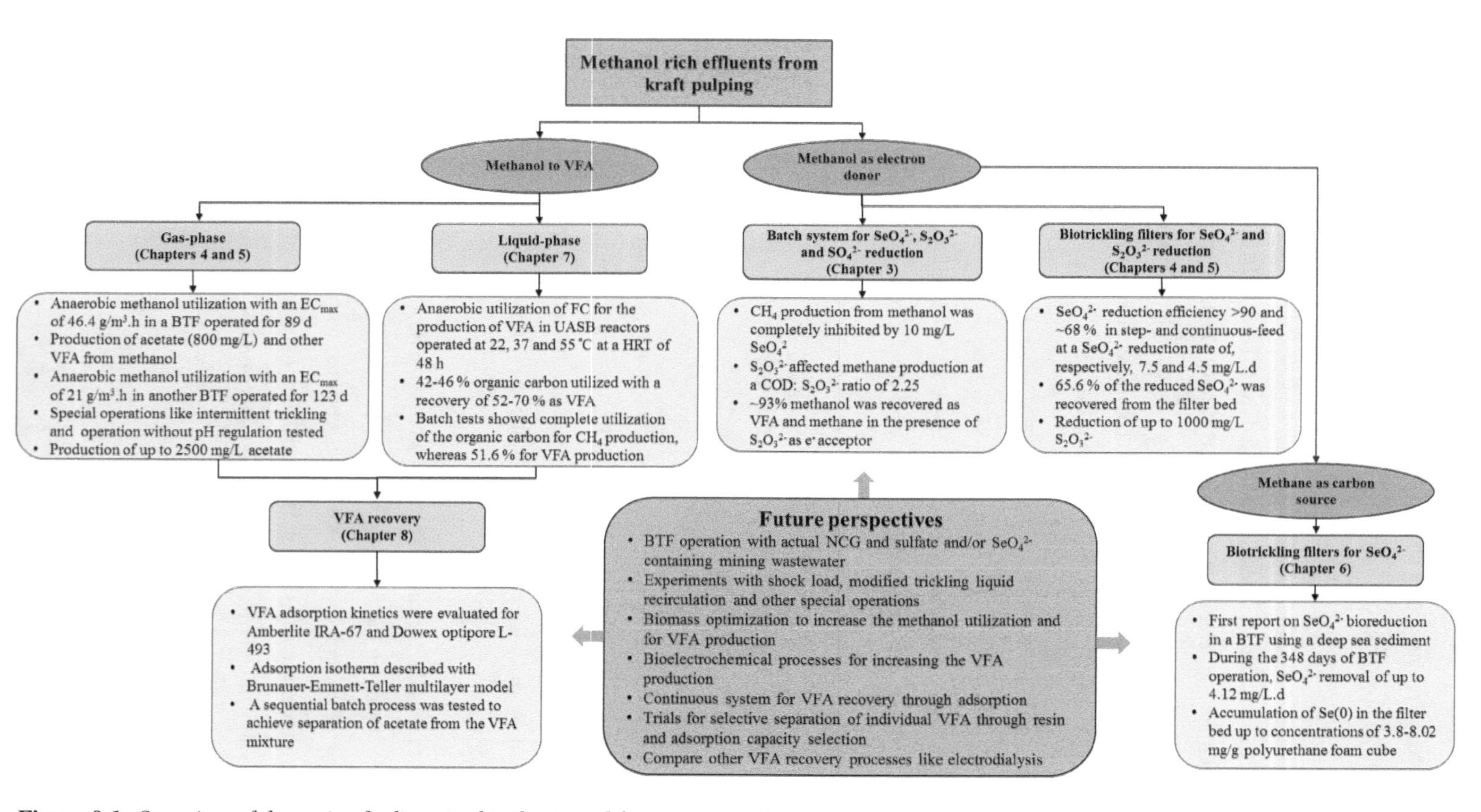

Figure 9.1: *Overview of the major findings in this thesis and future perspectives.*

9.2. VFA production from gas-phase methanol

Chapters 4 and 5 evaluated the anaerobic treatment of gas-phase methanol through its conversion to VFA. Gas-phase methanol utilization was tested for the first time in two anaerobic BTFs operated for 123 and 89 d, respectively, with thiosulfate and selenate rich trickling liquid. A maximum elimination capacity (E_{max}) of 21 and 46 $g/m^3.h$ was achieved, respectively, at a methanol loading of 25-30 and 50-56 $g/m^3.h$ (***Figures 4.2 and 5.4***). Operational parameters with respect to trickling liquid such as intermittent liquid trickling, unregulated pH control and continuous liquid trickling did not have a considerable effect on the removal efficiency (RE) of gas-phase methanol, whereas the VFA production rates were negatively affected. However, in the BTF operated in the presence of selenate (~50 mg/L as step-feed) in the liquid-phase, at a gas-phase methanol loading in the range of 30-50 $g/m^3.h$, up to 5000 mg/L methanol accumulated in the trickling liquid (***Figure 5.4***). Subsequently, the respective selenate feeding cycle was stopped and the following phase was started with fresh selenate rich trickling liquid (**Chapter 5**).
Overall, methanol utilization in the anaerobically operated BTF was > 95% at a methanol loading rate of 7-25 $g/m^3.h$ in the thiosulfate fed BTF (**Chapter 4**), while it was > 85% at a methanol loading rate of 7-25 $g/m^3.h$ in the selenate fed BTF (**Chapter 5**). Methane production from the anaerobic utilization of methanol was insignificant (< 5% *v/v*) and VFA (acetate, propionate, isobutyrate, butyrate, isovalerate and valerate) were the main end-products. A maximum of 2500 mg/L of acetate, 200 mg/L of propionate, 150 mg/L of isovalerate and 100 mg/L isobutyrate was produced in the BTF fed with thiosulfate and up to 800 mg/L of acetate and 117 mg/L of propionate in the BTF fed with selenate.

9.3. VFA production from liquid-phase methanol

Anaerobic treatment and VFA production from FC, a methanol-rich effluent collected from a Kraft pulping mill in Finland, were tested in three UASB reactors operated at 22, 37 and 55 °C. Organic carbon (methanol, ethanol and acetone) utilization of 42-46% was achieved and 52-70% of the utilized organic carbon was converted to VFA. Prior to acetogenesis of FC in the three UASB reactors, enrichment of acetogenic biomass was carried out for a period of 100 d (***Figure 7.1***). The operational parameters that are reported to inhibit the methanogenesis step and induce acetogenesis, such as the supply of a cobalt deprived trace element solution, pH shock, carbonate addition, increased organic loading rate (OLR), low pH (5.5) of the influent and addition of 2-bromoethanesulfonate (BES) were applied (***Chapter 7.2***). Along with the production of acetate, propionate, isobutyrate, butyrate, isovalerate and valerate were also produced in the three UASB reactors (***Figure** 7.2*). The presence of ethanol and acetone in the FC resulted in a higher percentage of C_4-C_5 VFA production compared with methanol rich synthetic wastewater. Furthermore, the results from batch activity tests to compare the prospects of methane versus VFA production showed complete removal of the organic carbon fraction through methane production, whereas 51.6% organic carbon utilization was achieved through VFA production (***Table** 7.2*). Thus, from a practical application perspective, methane production is a more suitable resource recovery approach for the treatment of FC compared to VFA production (**Chapter** 7).

9.4. Methanol rich effluents as carbon source for the reduction of S and Se oxyanions

Use of gas-phase methanol as energy source for the anaerobic reduction of selenate or thiosulfate rich synthetic wastewaters were tested in BTFs **(Chapter 4, 5)** after comparing the anaerobic utilization of methanol in the presence of selenate, sulfate and thiosulfate in batch studies **(Chapter 3)**. Concerning liquid phase effluents, several studies have reported the anaerobic utilization of methanol rich synthetic wastewaters for the reduction of S oxyanions (Vallero et al., 2004). Up to 1000 mg/L of thiosulfate **(Chapter 4)** and 50 mg/L of selenate **(Chapter 5)** were fed to the anaerobically operated BTFs in step feed mode and their reduction were monitored. Thiosulfate reduction under special operation conditions intermittent trickling, operation without pH control were demonstrated and a reduction rate of up to 370 mg/L.d was achieved in the BTF **(Chapter 4, Table 4.1)**. Selenate reduction under continuous and step feeding conditions were tested and average selenate reduction rates of 7.3 and 4.5 mg/L.d were achieved during step and continuous feeding conditions. Up to 65 % of the total Se fed as selenate was recovered from the filter bed **(Chapter 5)**.

9.5. VFA recovery using ion-exchange resins

Waste-derived VFA production is gaining a lot of attention due to the ease of production and scale-up using a wide range of organic substrates and a variety of mixed microbial consortia (Lee et al., 2014). Several physico-chemical processes such as electrodialysis, solvent-extraction, distillation, precipitation and adsorption have been tested for VFA recovery (Reyhanitash et al., 2016). As part of this thesis, adsorption onto anion exchange resins was examined for the recovery of VFA such as acetic, propionic, isobutyric, butyric, isovaleric and valeric acid in batch systems **(Chapter 8)**. Granular activated carbon and 11 anion exchange resins were initially screened for VFA adsorption (***Figure** 8.1*). Among them, Amberlite IRA-

67 and Dowex optipore L-493 were chosen for further investigation. The adsorption kinetics were evaluated using the pseudo-second-order and intraparticle diffusion model. The single layer isotherm models such as the Langmuir's single-component and multi-component system models were inadequate to describe the VFA adsorption mechanism based on the resin capacities for individual VFA (***Table 8.7***). Therefore, the Brunauer-Emmett-Teller (BET) multi-layer adsorption model was used. In a multi-component VFA system, the maximum adsorption capacity (q_m) of Amberlite IRA-67 and Dowex optipore L-493 varied considerably for the individual VFA (***Table 8.7***). Based on the selective adsorption capacity of the resins, this study explored the separation of an individual component from a mixture using a sequential adsorption process, specifically, acetate recovery from the VFA mixture.

9.6. Future perspectives

9.5.1. Treatment of methanol rich effluents

Figure 9.2 provides an overview of the application of the research carried out in this thesis for the management of gaseous emissions and liquid effluents of a chemical pulping mill. The methanol-rich NCG was anaerobically treated in BTF, with additional organic carbon recovery through VFA production. Furthermore, by using S- and Se-oxyanion rich industrial effluents in the trickling liquid, bioreduction of these oxyanions was achieved. To avoid production of NCG in the first place, the methanol-rich effluents such as FC that usually undergo further evaporation to recover reusable water, and eventually end up as the NCG, can be treated in UASB reactors. Using UASB reactors, the organic carbon can be recovered as VFA or methane, with an added advantage of avoiding CO_2, methanol and other VOC rich emissions into the atmosphere.

Anaerobic utilization of methanol, ethanol and acetone from FC and VFA production was examined in **Chapter** 7. However, the fate of reduced sulfur compounds present in the FC was not ascertained in this study. Furthermore, in **Chapters 4 and 5**, the study focused only on anaerobic gas-phase methanol utilization. Future studies should also focus on the anaerobic utilization of real NCG from a pulp mill, since it contains also other malodourous reduced sulfur compounds and volatile organic carbon such as methyl mercaptan (CH_3SH), dimethyl sulfide (CH_3SCH_3) and dimethyl disulfide (CH_3SSCH_3), and α-pinene (Suhr et al., 2015). Besides, it is important to examine the speciation of S compounds in bioreactors in order to understand their fate and effect on organic carbon conversion and VFA yields.

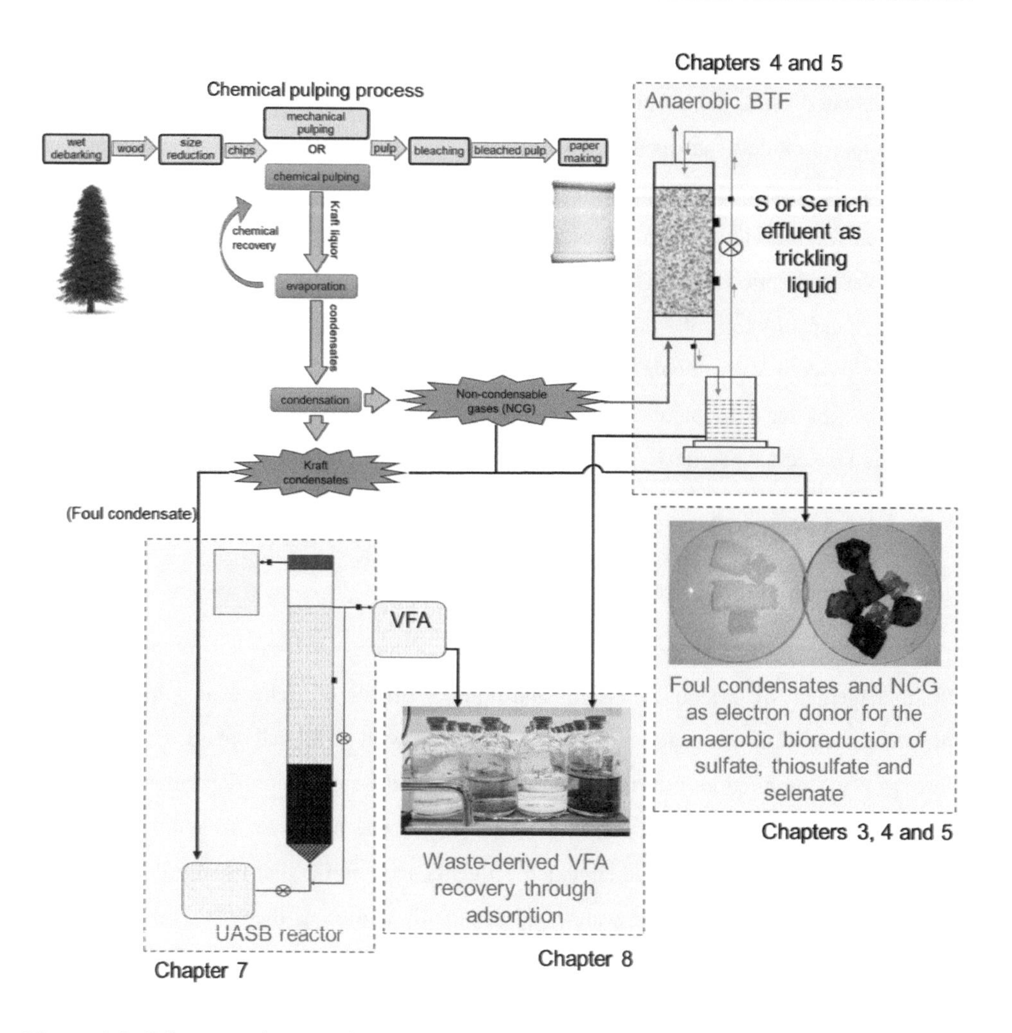

Figure 9.2: *Schematic showing the bioconversions of methanol- rich pulping industry effluents like non-condensable gases and foul condensates to VFA, and VFA recovery through adsorption.*

Methanol-rich effluents can be used as carbon source for the treatment of Se and S oxyanion rich industrial effluents. In order to further improve the performance of the bioreactors (BTF and/or UASB reactors), examination of variations in operational parameters such as HRT, OLR, and liquid recirculation rate is necessary. More research on the activity of the acetogenic biomass to optimise the VFA production and methanol removal at a higher OLR would be helpful to design high-rate VFA production bioreactors.

9.5.2. Methanol rich effluents as energy source for the reduction of S and Se oxyanion rich effluents

Anaerobic reduction of selenate and thiosulfate in synthetic wastewater using gas-phase methanol was demonstrated in Chapters 4 and 5. The presence of selenate or thiosulfate affected the methanol utilization by methanogens and the utilized methanol was converted to VFA **(Chapters 3, 4 and 5)**. Further investigation using effluents from selenium, gold and lead mining industries that contain up to 620, 33 and 7 mg/L Se oxyanions or effluents from petroleum refining industries that could contain up to 50 g/L thiosulfate is necessary. Furthermore, it is interesting to study the combined effect of the S or Se oxyanions together with the constituents of a real methanol rich gas phase effluent such as NCG on the biomass.it was clearly elucidated that the methanogens are inhibited by the presence of S or Se oxyanions **(Chapter 3)**, implying that the resource recovery is possible through VFA production.

9.5.3.Waste-derived VFA

Concerning the waste-derived VFA, although several studies have explored VFA production from different types of organic biomass and recovery through physical processes such as electrodialysis, adsorption, precipitation and solvent extraction, the recovery of individual VFA still remains a major challenge (Eregowda et al., 2018). **Chapter 8** showed a sequential batch process using different types of anion exchange resins is a promising candidate for the recovery of individual VFA. Furthermore, this approach can be applied for the separation of other ionic compounds such as SO_4^{2-}, SeO_4^{2-} and lactate. Production of VFA from methanol-rich effluents and recovery through adsorption on anion exchange resins was demonstrated in this thesis as a proof-of-concept. Further studies using continuously operated adsorption columns and optimization of the reactors hydrodynamic and operational parameters should be performed to assess the practical feasibility of scaling up the process for industrial applications and making it economically viable.

9.5.4. Gas-to-liquid fuel technologies

With an increase in the global methane reserves from 113 trillion m^3 in 1990 to 194 trillion m^3 in 2012 and an increasing trend towards the adaptation of anaerobic digestion for the treatment of organic waste substances, methane is an abundant reserve (Ge et al., 2014). Furthermore, due to the low methane content (in natural gas) and the presence of other impurities (in biogas), petroleum drilling and AD facilities often restore to flaring. The total flared gas volume in 2012 was estimated at 143 billion m^3, corresponding to 3.5% of the total global production (Elvidge et al., 2016).

As a potential solution for the flaring, several studies have suggested the conversion of gas-to-liquid fuels, wherein methane is converted to methanol, VFA and other value-added platform chemicals. Thermo-chemical conversions involving syngas as an intermediate is currently the most popular approach (Ge et al., 2014). However, few studies have attempted biological conversions, involving methanotrophs, to produce acetate, carboxylic acids and other chemicals (Garg et al., 2018; Mueller et al., 2015).

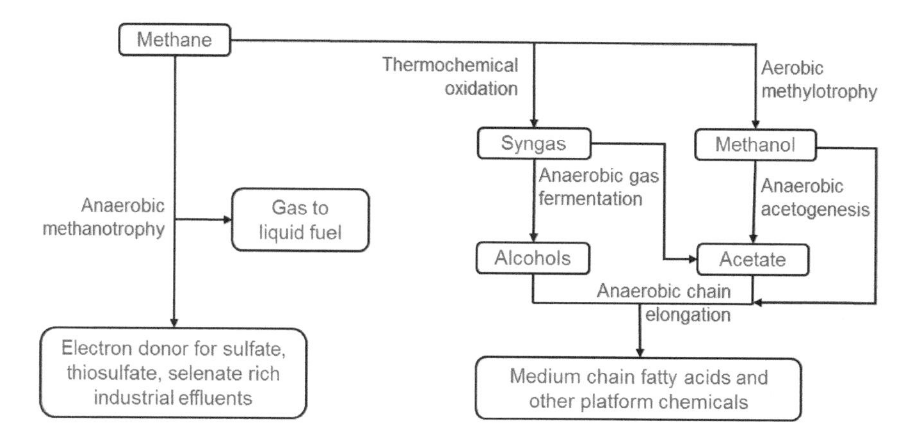

Figure 9.3: *Overview of the applications of the findings from this study for the development of indirect methane-to-liquid fuel technologies.*

The schematic for an alternative (indirect) approach through integrated bioprocesses involving different microorganisms to achieve the conversion of methane to methanol and subsequently, methanol to other liquid fuels (Fei et al., 2014; Ge et al., 2014; Yang et al., 2014) is shown in ***Figure 9.3***. The results from **Chapters 3, 4, 5 and** 7 can be applied to achieve indirect gas-to-liquid fuel conversions. A maximum production of up to 2500 mg/L acetate, 250 mg/L propionate, 50 mg/L isobutyrate, 102 mg/L butyrate, 150 mg/L isovalerate and 175 mg/L valerate from the anaerobic utilization of methanol-rich FC was shown in **Chapter** 7.

9.7. Conclusions

This thesis presents various hybrid approaches for the treatment and utilization of methanol-rich gaseous and liquid effluents of a chemical pulping mill for VFA production and recovery. Anaerobic biodegradation of gas-phase methanol was demonstrated in the BTFs together with

VFA production. Gas-phase methanol can be utilized as a carbon source and electron donor for the reduction of S and Se oxyanion rich wastewater. Methanol utilization for methanogenesis was completely inhibited in the BTF and in the presence of selenate at concentrations > 10 mg/L. Concerning the treatment of liquid effluents from the Kraft pulping process, e.g. FC, resource recovery in terms of methane production is economically and technically more viable over VFA production. In the case of gaseous effluents such as NCG, resource recovery in terms of biological methanol utilization through VFA production is preferable over methane production. As a downstream process, the recovery of individual VFA can be achieved through sequential adsorption using different types of ion exchange resins.

References

Bajpai, P., 2016. Pulp and Paper Industry: Energy Conservation. Elsevier, Amsterdam, Netherlands.

Baumann, I., Westermann, P., 2016. Microbial production of short chain fatty acids from lignocellulosic biomass: current processes and market. Biomed Res. Int. 2016.

CEPI, 2017. Key Statistics, European pulp and paper industry.

Chen, W.S., Ye, Y., Steinbusch, K.J.J., Strik, D.P.B.T.B., Buisman, C.J.N., 2016. Methanol as an alternative electron donor in chain elongation for butyrate and caproate formation. Biomass and Bioenergy. 93, 201–208.

Coma, M., Vilchez-Vargas, R., Roume, H., Jauregui, R., Pieper, D.H., Rabaey, K., 2016. Product diversity linked to substrate usage in chain elongation by mixed-culture fermentation. Environ. Sci. Technol. 50, 6467–6476.

Da Silva, A.H., Miranda, E.A., 2013. Adsorption/desorption of organic acids onto different adsorbents for their recovery from fermentation broths. J. Chem. Eng. Data. 58, 1454–1463.

Elvidge, C.D., Zhizhin, M., Baugh, K., Hsu, F.C., Ghosh, T., 2016. Methods for global survey of natural gas flaring from visible infrared imaging radiometer suite data. Energies. 9, 1-16.

Eregowda, T., Matanhike, L., Rene, E.R., Lens, P.N.L., 2018. Performance of a biotrickling filter for anaerobic utilization of gas-phase methanol coupled to thiosulphate reduction and resource recovery through volatile fatty acids production. Bioresour. Technol. 263, 591–600.

Fei, Q., Guarnieri, M.T., Tao, L., Laurens, L.M.L., Dowe, N., Pienkos, P.T., 2014. Bioconversion of natural gas to liquid fuel: Opportunities and challenges. Biotechnol. Adv. 32, 596–614.

Garg, S., Wu, H., Clomburg, J.M., Bennett, G.N., 2018. Bioconversion of methane to C-4 carboxylic acids using carbon flux through acetyl-CoA in engineered *Methylomicrobium buryatense* 5GB1C. Metabol. Eng. 48, 175–183.

Ge, X., Yang, L., Sheets, J.P., Yu, Z., Li, Y., 2014. Biological conversion of methane to liquid fuels: status and opportunities. Biotechnol. Adv. 32, 1460–1475.

Lee, W.S., Chua, A.S.M., Yeoh, H.K., Ngoh, G.C., 2014. A review of the production and applications of waste-derived volatile fatty acids. Chem. Eng. J. 235, 83–99.

Mueller, T.J., Grisewood, M.J., Nazem-Bokaee, H., Gopalakrishnan, S., Ferry, J.G., Wood, T.K., Maranas, C.D., 2015. Methane oxidation by anaerobic archaea for conversion to liquid fuels. J. Ind. Microbiol. Biotechnol. 42, 391–401.

Rebecchi, S., Pinelli, D., Bertin, L., Zama, F., Fava, F., Frascari, D., 2016. Volatile fatty acids recovery from the effluent of an acidogenic digestion process fed with grape pomace by adsorption on ion exchange resins. Chem. Eng. J. 306, 629–639.

Reyhanitash, E., Zaalberg, B., Kersten, S.R.A., Schuur, B., 2016. Extraction of volatile fatty acids from fermented wastewater. Sep. Purif. Technol. 161, 61–68.

Roghair, M., Hoogstad, T., Strik, D.P.B.T.B., Plugge, C.M., Timmers, P.H.A., Weusthuis, R.A., Bruins, M.E., Buisman, C.J.N., 2018. Controlling ethanol use in chain elongation by CO_2 loading rate. Environ. Sci. Technol. 52, 1496–1506.

Suhr, M., Klein, G., Kourti, I., Gonzalo, M.R., Santonja, G.G., Roudier, S., Sancho, L.D., 2015. Best available techniques (BAT) – reference document for the production of pulp, paper and board. Eur. Comm. 1–906.

TRI Program, 2017. 2017 TRI Factsheet for Chemical METHANOL.

Yang, L., Ge, X., Wan, C., Yu, F., Li, Y., 2014. Progress and perspectives in converting biogas to transportation fuels. Renew. Sustain. Energy Rev. 40, 1133–115

Curriculum vitae

Tejaswini Eregowda

193/21, 1st left cross, Girinager 4th phase, 560085 Bengaluru (India)

+917483958600

tejaswini.gowda.me@gmail.com

WORK EXPERIENCE

Nov 2015 – May 2019	PhD fellow Department of Environmental Engineering and Water Technology, IHE Delft - Institute for Water Education, Delft (Netherlands)
Nov 2014 – Oct 2015	Informantion Officer ENVIS, Environmental Management and Policy Research Institute, Bangalore (India)
Aug 2013 – Feb 2014	Production Assistant Gentech Propagation Ltd., Dundee (United Kingdom)

EDUCATION AND TRAINING

Sep 2007 – Jun 2011	Bachelor of Engineering in Biotechnology PES Institute of Technology, Bangalore (India) First class with distinction	EQF level 6
Nov 2011 – Jun 2013	Master of Technology in Environmental Engineering Manipal Institute of Technology, Manipal (India) First class with distinction	EQF level 7
May 2012 – Apr 2013	Project for M.Tech. Dissertation TERI-SRC, Bangalore (India) Project title: Small scale municipal wastewater treatment using sphagnum moss.	
Feb 2011 – Apr 2011	Project for B.E. Dissertation Aristogene Biosciences pvt. ltd., Bangalore (India) Project title: Development of standard method for quantification of arsenic in the water sample.	
Jul 2010 – Aug 2010	Training Syngene international ltd., Biocon, Bengaluru (India) Training on the software module Mastero-by Schrodinger for analysis of active sites of given enzymes, designing and validating agonistic and antagonistic compounds.	

Jun 2010 – Sep 2010 Project

University of Agricultural Sciences, GKVK, Bengaluru (India)

Molecular characterization of *Bacillus megaterium* isolated from different Agro climatic zones of Karnataka and its effects on *Eupatorium triplinerve* and *Mentha piperita.*

Jun 2009 – Aug 2009 Project

Best Biotek Research Labs (P) Ltd, Bengaluru (India)

Project title: Extraction, purification, kinetics, immobilization, preservation of the enzyme bromelain.

ADDITIONAL INFORMATION

Funding

- Recipient of AICTE-GATE Scholarship during M.Tech degree.
- Recipient of PhD fellowship under Marie Skłodowska-Curie European Joint doctorate (EJD) in Advanced Biological Waste-To-Energy Technologies (ABWET) funded from Horizon 2020.

Publications

- Eregowda, T., Rene, E.R., Rintala, J., Lens, P.N.L., 2019. Volatile fatty acid adsorption on anion exchange resins: kinetics and selective recovery of acetic acid. Sep. Sci. Technol. 1-13.
- Eregowda, T., Rene, E.R., Lens, P.N.L., 2019. Gas-phase methanol fed anaerobic biotrickling filter for the reduction of selenate under step and continuous feeding conditions. Chemosphere. 225, 406-413.
- Eregowda, T., Matanhike, L., Rene, E.R., Lens, P.N.L., 2018. Performance of a biotrickling filter for anaerobic utilization of gas-phase methanol coupled to thiosulphate reduction and resource recovery through volatile fatty acids production. Bioresour. Technol. 263, 591–600.
- Sandeep, C, Thejas, M.S., Patra, S, Gowda, T, Raman, R.V., Radhika, M, Suresh, C.K, Mulla, SR (2011) Growth response of ayapana on inoculation with *Bacillus megaterium* isolated from different soil types of various agroclimatic zones of Karnataka. J. Phytol. 3, 13–18.
- Sandeep, C., Raman, RV, Radhika, M., Thejas, MS, Patra, S., Gowda, T., Suresh, C.K., Mulla, SR (2011). Effect of inoculation of *Bacillus megaterium* isolates on growth, biomass and nutrient content of Peppermint. *J. Phytol. 3*, 19-24.

Submitted manuscript:

- Eregowda, T., Kokko, M.E., Rene, E.R., Rintala, J., Lens, P.N.L., 2019. VFA production from Kraft mill foul condensate in an upflow anaerobic sludge blanket (UASB) reactor. Manuscript submitted to J. Hazard. Mater.
- Eregowda, T., Rene, E.R., Matanhike, L., Lens, P.N.L., 2019. Effect of selenate, thiosulfate on VFA and methane production through anaerobic methylotrophy. Manuscript submitted to Environ. Sci. Pollut. Res.

Conferences

- Eregowda, T., Rene, E.R., Kokko, M., Rintala, J., Lens, P.N.L., Resource recovery from methanol rich industrial gaseous and liquid effluents through acetogenesis, G16 and ABWET doctorate project final conference 2018, Napoli, Italy.
- Eregowda, T., Rene, E.R., Kokko, M., Rintala, J., Lens, P.N.L., Acetogenesis of foul condensate: biomass enrichment and treatment in upflow anaerobic sludge blanket (UASB) reactor, PhD symposium 'nature for water' 2018, IHE-Delft, Delft the Netherlands.
- Eregowda, T., Rene, E.R., Wadgaonkar, S.L., Vossenberg, J., Lens, P.N.L., Selenate bioreduction in the presence of methane using slow growing biomass from deep sea sediment in a biotrickling filter, 7th International conference on biotechniques for air pollution control and bioenergy (Biotechniques-2017), La Coruña, Spain
- Eregowda, T., Rene, E.R., Vossenberg, J., Lens, P.N.L., Bioreduction of selenate using methane as electron donor in a biotrickling filter, 1st ABWET international conference, waste-to-bioenergy: applications in urban areas 2017, Paris, France.
- Eregowda, T., Mahendra, N., Ganesha, A., Babu, S.M., Waste to watts production of electricity from materials that have lost their economic value, Int.Green Campus Summit 2013, Pondicherry, India.
- Eregowda, T., Radhika, M., Sabath, S., Cloning and expression studies of arsenate reductase gene from *Pseudomonas aeruginosa* (Pa01) for detection of arsenic in contaminated water, Bangalore INDIA BIO 2011 Int. Conf., Bengaluru, India.

Netherlands Research School for the
Socio-Economic and Natural Sciences of the Environment

D I P L O M A

For specialised PhD training

The Netherlands Research School for the
Socio-Economic and Natural Sciences of the Environment
(SENSE) declares that

Tejaswini Eregowda

born on 19 October 1989 in Holenarsipura, India

has successfully fulfilled all requirements of the
Educational Programme of SENSE.

Tampere, 23 May 2019

The Chairman of the SENSE board

Prof. dr. Martin Wassen

the SENSE Director of Education

Dr. Ad van Dommelen

The SENSE Research School has been accredited by the Royal Netherlands Academy of Arts and Sciences (KNAW)

K O N I N K L I J K E N E D E R L A N D S E
A K A D E M I E V A N W E T E N S C H A P P E N

The SENSE Research School declares that Tejaswini Eregowda has successfully fulfilled all requirements of the Educational PhD Programme of SENSE with a work load of 62.9 EC, including the following activities:

SENSE PhD Courses

- Environmental research in context (2017)
- Research in context activity: 'Co-organizing ETeCOS3 and ABWET PhD Summer School (23-27 May 2016, Delft) and COST Training School (25-27 May 2016, Delft), as well as preparing Abstracts Book'

Other PhD and Advanced MSc Courses

- Orientation to doctoral studies, Tampere University of Technology (2016)
- Research writing, Tampere University of Technology 2016)
- Research ethics, Tampere University of Technology (2016)
- Research paper writing, TUT (2018)
- Biological wastewater treatment: principles, IHE Delft (2017)
- Biogas technology, Tampere University of Technology (2018)
- Webinar on frontiers in microbiology, Tampere University of Technology (2017)
- Summer course on mathematical modelling, University of Cassino (2017)
- Introductory course on advance biological waste-to-energy technologies, IHE Delft (2016)
- ETeCoS3 and 2nd ABWET summer school, on Contaminated Sediments -characterization and remediation, IHE Delft, Netherlands (2016)
- 3rd summer school on biological treatment of solid waste, University of Gaeta (2017)
- 4th ABWET summer school, Tampere University of Technology (2018)

Management and Didactic Skills Training

- Contributed towards the making of the video lecture on 'sulfate reduction activity of an activated sludge is estimated by applying a sacrificial bottle test', part of "Online courses on experimental methods in wastewater treatment (2016).
- Supervising MSc student with thesis entitled 'Anaerobic biotrickling filter for the removal of volatile inorganic and organic compounds' (2017)
- Supervising Erasmus student (2017)

Oral Presentations

- *Resource recovery from methanol rich industrial gaseous and liquid effluents through acetogenesis*. G16-ABWET conference, 6-7 December, Naples, Italy

SENSE Coordinator PhD Education

Dr. Peter Vermeulen